Rapid Load Testing on Piles

Rapid Load Testing on Piles

Interpretation Guidelines

CUR publication 230

Paul Hölscher
Deltares, Delft, The Netherlands

Henk Brassinga
City of Rotterdam, The Netherlands

Michael Brown
Dundee University, Dundee, UK

Peter Middendorp
*Allnamics Pile Testing Experts BV, Voorburg,
The Netherlands*

Maarten Profittlich
*Fugro Onshore Geotechnics, Leidschendam,
The Netherlands*

Frits van Tol
*Delft University of Technology, Delft, The Netherlands
Department of Civil Engineering and Geosciences,
Deltares, Delft, The Netherlands*

CRC Press
Taylor & Francis Group
Boca Raton London New York

CRC Press is an imprint of the
Taylor & Francis Group, an **informa** business

A BALKEMA BOOK

CUR

BUILDING & INFRASTRUCTURE
CUR guidelines 230

CRC Press
Taylor & Francis Group
6000 Broken Sound Parkway NW, Suite 300
Boca Raton, FL 33487-2742

First issued in paperback 2018

CRC Press/Balkema is an imprint of the Taylor & Francis Group, an informa business

ISBN-13: 978-0-415-69520-6 (hbk)
ISBN-13: 978-1-138-11377-0 (pbk)

Library of Congress Cataloging-in-Publication Data

Rapid load testing on piles : interpretation guidelines / Paul Hölscher ... [et al.].
p. cm.
Summary: "A Rapid Load Test (RLT) developed to determine the initial stiffness and bearing capacity, is an economical and practical alternative for a Static Load Test (SLT) . However, the application of RLT used to be hampered by uncertainty about the interpretation of the test results. These guidelines attempt to offer clear guidance on the available analysis techniques and their reliability"-- Provided by publisher.
Includes bibliographical references and index.
ISBN 978-0-415-69520-6 (hardback)
1. Piling (Civil engineering)--Materials--Dynamic testing. I. Hölscher, Paul, 1960- II. Title.

TA780.R35 2011
624.1'54--dc23
2011029850

Visit the Taylor & Francis Web site at
http://www.taylorandfrancis.com

and the CRC Press Web site at
http://www.crcpress.com

Contents

Preface

The current Guidelines on the Interpretation of Rapid Load Testing on Piles is the final deliverable of the Dutch Delft Cluster knowledge program "New perspective on foundations and building excavations", which is part of the research topic "Better controlled use of the underground".

In the research program, conducted by Deltares together with Delft University of Technology, the need for affordable and reliable methods for measuring the actual load capacity of piles was examined, and as a first step a draft standard, which fits the framework of Eurocode 7, was developed. This Standard is now under final review for formal approval.

The results of the research were then translated into a practical guideline by a CUR/Delft Cluster committee between 2008 and 2011. This CUR/Delft Cluster committee was established with the participation of a large number of stakeholders and at the time this report is published consists of the following individuals:

Prof. A.F. van Tol, chairman	Deltares/Delft University of Technology
Dr. P. Hölscher, reporter	Deltares
H.E. Brassinga, secretary	City of Rotterdam
P. Middendorp M.Sc	Allnamics Pile Testing Experts BV
P. Anemaat	COBc, city of Vlaardingen
M.J. Profittlich	Fugro
P.P.T. Litjens	Rijkswaterstaat/Ministry of Infrastructure and the Environment
A. Ramkema	VWS Geotechniek
R. van Foeken	IHC Hydrohammer
E. Revoort	Verstraeten BV
N. Goedhart	Ballast Nedam
C.J. Vroom	Vroom Funderingstechnieken
T. Siemerink, coordinator	CUR Building & Infrastructure

The committee would like to thank the following organisations for their financial support of the project:

- Rijkswaterstaat Dienst Infrastructuur
- Shell Global Solutions International BV
- IHC Hydrohammer

- VWS Geotechniek
- Ballast Nedam
- Berminghammer Inc. (Canada)

The committee would like to thank for their contribution to the execution of the field test in Waddinxveen:

- Profound BV
- Verstraeten BV
- City of Rotterdam
- Ballast Nedam

Prof. Matsumoto from Kanazawa University (Japan) is acknowledged for his contribution to the appendix.

Delft Cluster and CUR express their gratitude to all committee members and their organizations for their contribution and the results achieved by the CUR/Delft Cluster committee H410 Rapid Load Testing.

Figure 1 4 MN rapid load test device in the Netherlands.
Copyright: Allnamics Pile Testing Experts BV and Profound BV.

Summary

A Rapid Load Test (RLT) is an economical and practical alternative to a Static Load Test (SLT) for the determination of the initial stiffness and bearing capacity of piles. However, the application of RLT is hampered by uncertainty about the interpretation of the test results. These guidelines attempt to offer clear guidance on the available analysis techniques and their reliability.

These guidelines are related to an international standard which is drawn up by CEN.

These guidelines can be used in two ways:

1   Straight forward interpretation of test results, see chapters 1–3.
2   Interpretation with additional background information about the possibilities and limitations, see chapters 4–8.

Chapters 1 and 2 deal with the execution of the test and the presentation of the results. Chapter 3 gives a general overview of existing interpretation methods. This chapter refers to step-by-step descriptions of the two recommended interpretation methods.

The aspects which are important for the interpretation are discussed in chapter 4. It turns out that the inertia of the pile should be taken into account. In clay the rate effect (the dependency of strength and stiffness on loading rate) must also be taken into account. In sand and silt the generation of pore water pressures during an RLT plays a role. The velocity of the RLT is such, that the reaction of sand and silt might be considered as partially drained. These guidelines indicate how these effect can be compensated to obtain the static resistance in the final results.

Chapter 5 presents two interpretation methods:

–   A method for piles in sand, gravel, silt and piles on rock. This method is a simplified version of the Unloading Point Method (UPM), originally developed in the Netherlands.
–   A method for piles in clay. This method is a simplified version of the Sheffield Method (SHM) originally developed in the U.K.

This chapter gives a precise description of both methods combined with an example. For practical use, the methods are described in a step-by-step scheme in an Appendix of these guidelines.

The remaining part of these guidelines are concerned with aspects of RLT and the use of RLT test results:

– the available data which can be used for further validation of the interpretation methods;
– the recommended partial factors that fit within the framework of Eurocode 7 part 1;
– the influence of special situations, such as open ended piles and installation effects.

Chapter 1

Introduction and scope of the guideline

The Rapid Load Test (RLT) appears to be a good and economical alternative to the Static Load Test (SLT) on piles. Examples of this test method are Statnamic (Janes, Bermingham et al. 1991), the Pseudo-Static Pile Load Tester (Schellingerhout and Revoort 1996) and the Spring Hammer Device (Matsumoto, Matsuzawa et al. 2008). However, in Europe the application of such tests is hampered by the uncertainty about the interpretation of the test. This hindrance can be overcome by proper regulation of the test and its interpretation. These guidelines are meant to contribute to the quality of interpretation.

Currently, there is no generally accepted and applicable interpretation method for the results of an RLT. These guidelines give two well-defined methods for the practical interpretation of test results, including the theoretical background of these methods. Once the interpretation methods are defined, the results can be compared with the results of SLT's. In the longer term, this will give important information on the applicability of the interpretation methods and their reliability.

The rapid load test is applicable for:

– Checking the initial stiffness and bearing capacity of test and working piles.
– Acceptance of new pile types (trial piles) and/or construction techniques (preliminary pile).
– Judgment on the reusability of existing piles.

These guidelines are intended for two types of users:

1 Users who aim to obtain results without thorough understanding of the theoretical background of the interpretation methods.
2 Users who are interested in background information and knowledge about the possibilities and limitations.

The first type of user is recommended to pay attention to the Chapters 1 to 3 and to proceed with the step by step examples in appendices A and D.

These guidelines will explain two interpretation methods:

– A method for piles in sand, gravel, silt and piles on rock. This method is a simplified version of the Unloading Point Method (UPM), originally developed in the Netherlands.

– A method for piles in clay. This method is a simplified version of the Sheffield Method (SHM) originally developed in the U.K.

The recommendations of the CUR committee are:

– The stiffness calculated with UPM can be directly compared with the stiffness of static load tests for all soil types. No corrections or correlations with static load tests are required.
– For capacity determination UPM can be applied for piles in sand, gravel, silt and piles on rock. UPM needs the introduction of a reduction factor to compensate for loading rate effects. Based on database results this reduction is 5 to 10%.
– SHM can be applied for piles in clays. The soil dependant model parameters α and β have to be obtained from correlations with a static load test or from laboratory results.

Figure 1.1 The use of rapid pile tests can save time during the building process.
Copyright: Allnamics Pile Testing Experts BV and Profound BV.

Chapter 2

Test execution and test results

For quality-control, the execution of the test must satisfy the requirements in the latest version of the Standard for Rapid Load Testing (Appendix I).

For acceptance of new pile types and/or construction techniques, the instrumentation of the piles must comply with the requirements summarized in the guidelines and standards for static load tests. The requirements for the additional transducers must satisfy at least the requirement with respect to accuracy and sample frequency as mentioned in Section 4 of the Standard for RLT.

All measurements of additional transducers must be available in chart and digitally in ASCII format, in accordance with Section 6 of the Standard for RLT.

The results of the measurements must be presented as defined in the Standard.

The rapid load-displacement diagram has to be constructed. All relevant analysis parameters, pile information, soil information and site information in accordance with the RLT standard should be listed.

Chapter 3

General overview
of interpretation methods

3.1 INTRODUCTION

This chapter is meant to give the reader the basic knowledge in order to apply the interpretation methods and understand the following chapters.

Two interpretation methods will be presented:

1   For piles in sand, silt and piles on rock (Unloading Point Method, UPM).
2   For piles in clay (Sheffield Method, SHM).

A step by step approach to obtain derived static results for both interpretation methods is presented in Appendix A and D.

3.2 RELEVANT ASPECTS FOR THE INTERPRETATION

The following influences on RLT results are treated in this chapter:

– Influence of the inertia of the pile and the soil (3.2.1).
– Influence of the rate effect: pore water pressures and constitutive rate effects in clay.

Both interpretation methods include compensation for these influences in the test results.

Some special effects will be treated in Chapter 7:

– Influence/treatment of soil type (including layered soil).
– The treatment of open ended piles (including the plugging effect).

3.2.1 Inertia effects

During a rapid load test, the pile will be accelerated and gain some velocity. Finally the pile movement is stopped by the soil and the pile is decelerated. Acceleration and deceleration of the pile mass introduce inertia forces which have to be taken into account. So a rapid load test is not a static event: the influence of inertia must be considered.

The main difference between an RLT and a dynamic load test (DLT) is the duration of the loading. In a dynamic test the duration of loading is that short, that the load is removed from the pile head before or shortly after the moment that the stress wave reflection from the pile toe reaches the pile head. The duration of loading in a rapid load test, is so much longer that pile and soil behave in a quasi static way. Consequently, the stress wave phenomena in the pile may be neglected during a rapid load test but the inertia forces must be considered. In the interpretation methods in these guidelines, the pile is considered as a concentrated mass in combination with a spring to model the pile elasticity.

3.2.2   Rate effects

In these guidelines, the term "rate effect" is used to describe the dependency of the pile behaviour on the rapidity of loading. In sand, the rate effect is caused by the generation of pore water pressures, while in clay it is a constitutive property.

3.2.2.1   Pore water pressure

Excess pore water pressure plays an important role when the soil is undrained or partly drained during loading. For clay, the behaviour under a rapid load test can be considered as undrained. In general, silt and fine grained sand behave as partially drained material. In these cases the effect of pore water pressures must be handled with care. For coarse sand and gravel, the behaviour can be considered as fully drained. In this case, the excess pore water pressure does not play a role.

3.2.2.2   Constitutive rate effects

The constitutive rate effects are mainly observed in clay; for sand and silt, these effects are less important. For clay, an increase in strength with the rate of loading is observed in laboratory tests. When a pile in clay is tested with a high loading rate (compared with the 'static' rate), the increase in strength due to the loading rate must be considered.

3.3   DESCRIPTION OF INTERPRETATION METHODS

3.3.1   General

This section describes the methods which can be used to find the derived load-displacement curve from an RLT. This load-displacement curve is called derived, since it is based on the results of an RLT. Ideally, it should give a good comparison with the load-displacement diagram based on an SLT.

The following data must be measured:

- The force on the pile head – F(t).
- The acceleration of the pile head – a(t).
- The displacement of the pile head – w(t).

Additionally, the pile mass (its weight divided by the acceleration due to gravity, which is 9.81 m/s²) and the soil type must be known. Depending on the soil type, a method can be chosen, and then some additional data is required. The additional data depends on the chosen interpretation method.

In general, the interpretation method consists of the following steps:

3.3.1.1   Calculation of the soil resistance

This soil resistance is calculated from the measured force on the pile head and the inertial force of the pile: the pile mass multiplied by the measured acceleration.

$$F_{soil}(t) = F(t) - F_{inertia}(t) = F(t) - ma(t) \tag{3.1}$$

Where:

$F_{soil}(t)$ Force from the soil on the pile [N].
$F(t)$ (Measured) load on the pile head [N].
$F_{inertia}(t)$ Inertial force [N].
m   Mass of the pile [kg].
$a(t)$ (Measured) acceleration of the pile [m/s²].

3.3.1.2   Correction for loading rate effect

The soil force is corrected for the loading rate effects. In this step, the UPM and SHM methods differ significantly.

3.3.1.3   Construction of the derived static load-displacement

Indication of initial pile stiffness and maximum mobilised static resistance.

If the load on the pile head meets the requirements in the Standard, the pile is fully under compression and all parts of the pile move in similar directions. Then, it is acceptable to model the pile as a single mass and a spring and the acceleration measured at the pile head is valid for the pile as a whole. The force that corrects for inertia can be calculated from the multiplication of pile mass and measured acceleration.

– It should be taken into account that the influence of creep in the soil is not measured with an RLT.
– Finally the following aspect must be taken into account. The static behaviour of a pile during initial loading, unloading and reloading varies. Depending on the available equipment and preferences, the load can be applied with one loading cycle (single step) or with multiple cycles with increasing loading peaks (multiple steps). The final result is a derived static load-displacement curve for the test situation. So if the test is an initial load test on a virgin pile, the result reflects the behaviour of a virgin pile, and if the test includes an additional loading cycle or cycles to a static load test, the result reflects the behaviour at reloading. Therefore, it must be clearly defined whether the test includes an initial load test or a reload test.

Chapter 5 presents a more detailed and mathematical based discussion of the methods.

3.3.2   Choice of method

The methods are not always applicable. Table 3.1 gives an overview of expected applicability. Distinction is made between the determination of the ultimate bearing capacity and the derivation of the initial stiffness.

In table 3.1 it is assumed that the soil is homogeneous. For layered soils, documented data are available, but not elaborated in full detail to a practical model. For difficult situations, local experience for the interpretation is essential.

The stiffness calculated with UPM can be directly compared with the stiffness of static load tests. No corrections or correlations with static load tests are required.

Table 3.1 Overview of applicability of available interpretation methods.

Method	Property	clay	silt	sand	rock
UPM	Bearing capacity	+−	++	++	++
	Initial stiffness	++	++	++	++
SHM	Bearing capacity	++	+	−	−−
	Initial stiffness	+	+	−	−−

++ very suitable, + suitable, + − questionable,
− not really suitable , − − not suitable

3.3.3   Short description of the Unloading Point Method

The Unloading Point Method (UPM) starts the interpretation by finding the so-called unloading point (maximum displacement). At this point, the velocity of the pile is zero.

The static load on the pile head is obtained by correcting the rapid load on the pile head for the inertia effect (pile mass × acceleration) at the unloading point.

Then this force is corrected for rate effects using an empirical factor η, which is based on the comparison of RLT and SLT (database) results, see table 5.1.

It is assumed that the soil resistance at the unloading point corrected for inertia and loading rate, is a point on the static load-displacement diagram. It should be noted that for the Unloading Point Method until now all calculations are carried out at one point only.

The stiffness is determined from the displacement under characteristic load (working load). The method to find the load-displacement curve depends on the number of cycles in the test (number of RLT's on the pile). For a single cycle test, the displacement under characteristic load is used, for a multi-cycle test, the procedure uses the cycle which has the peak load close to the characteristic load.

Finally, the full curve can be drawn. For a single cycle test, a hyperbolic load-displacement diagram may be assumed. For a multi-cycle test the (corrected) unloading points for all cycles can be combined by a best fit curve or a best fit hyperbolic load-displacement diagram may be constructed.

This method is described in detail in Appendix A, B and C.

3.3.4   Short description of the Sheffield Method

The Sheffield Method uses the full time signals (all data points/time steps) for the calculation. Therefore, the analysis by this method will be performed numerically.

The static load on the pile head is obtained by correcting the rapid load on the pile head for the inertia effect (pile mass × acceleration) for all time steps. This force is corrected for rate effects using numerical factors α, β, which are calculated from the velocity of the pile as a function of time, using the rate factors for the clay. The rate factors α, β define the dependency of the clay behaviour as a function of pile velocity. These factors are preferably determined by laboratory tests, but practical values are available. The velocity can be calculated from the measured acceleration and/or measured displacement.

Since this procedure is carried out for all time steps in the measurement, the full load-displacement diagram is available directly.

The calculation procedure is not valid in the full range of the test. The proposed calculation method is only applicable to the downward motion. This method is described in detail in Appendix D.

3.3.5   Methods with more degrees of freedom

More elaborate calculation methods are available, such as 1-D wave equation analysis and 3-D finite element analysis. These methods are more complicated and the selection of proper soil model parameters requires additional information and more expertise. This introduces subjective elements in the analysis. These methods can however provide more insight in the results of RLT on piles for the assumed soil model parameters.

The 1-D method is based on signal matching. It is generally applied to the analysis of dynamic load testing (DLT) [e.g. CAPWAP, TNOWAVE and DLTWAVE]. The essential difference between the DLT and RLT is the duration of the loading on the pile head and as a consequence the length of the wave in the pile. Because of the relative low loading rise time and the duration of the loading RLT does not offer the possibility to distinguish clearly the individual response of each soil layer. This can be understood from signal processing theory: the frequency content of the signal and the associated resolution is too low to detect such differences.

More practically, it can be assumed that during an RLT all cross-sections of the pile move in the same velocity range. This allows the modelling of the pile as a concentrated mass with springs, representing the pile stiffness.

In 3-D finite element analysis [e.g. PLAXIS] the RLT can be simulated, to obtain a simulated RLT that fits the measurements, as well as obtaining a more realistic static load-displacement curve. Because with this method an "arbitrary" estimate is made with respect to the damping, additional information (e.g. static tests) is required to obtain a reliable interpretation (see e.g. Horikoshi, Kato et al. 1998).

It should be recalled that the results in this section focus on the interpretation of measurements at the pile head during an RLT (back calculation). The 1-D methods (GRLWEAP, PDPWAVE) and 3-D methods are of course applicable to simulate the tests, based on known properties of pile, soil and load (forward calculation). Generally, the standard models for dynamic calculations in both 1-D and 3-D do not incorporate the rate effects in soil. Before using 1-D or 3-D models, it should be evaluated that the relevant rate effects are included in the constitutive modelling.

As always, when using either of these (or any other) method for the interpretation of RLT tests outside of the area of experience (e.g. pile size, installation method, soil type or displacement range) where a sufficient volume of experience exists, additional information is required to validate the results. This could for instance result in the requirement for a static load test to correlate the RLT results.

3.4 SAFETY APPROACH

Considering the partial factors, two aspects must be taken into account:

- The uncertainty due to the testing method: a rapid load test is not a static test; due to the interpretation method, the real static stiffness or bearing capacity for a tested pile might differ from the value found from a static test.
- The uncertainty due to the spatial spreading of the properties of soil (and piles). In order to judge the safety of a full foundation, the number of tested piles with respect to the total number of piles should be taken into account.

Figure 3.1 Execution of a 16 MN rapid load test.
Copyright: Fugro and VWS Geotechniek BV.

Chapter 4

Theoretical aspects

4.1   INTRODUCTION

In this chapter, the most relevant aspects which must be taken into account are discussed theoretically. While interpreting the results of an RLT, one should be aware of these aspects. This chapter doesn't pretend to present a scientific complete discussion.

Two important aspects are discussed:

– Dynamic effects.
– Rate effects.

4.2   THE INVERSE PROBLEM

A procedure to calculate the derived static load displacement curve from the measured data can be written generally as a function of the measured data and known parameters:

$$F_{stat,der}(t) = g[F(t), w(t), v(t), a(t); m]$$ (4.1)

Where:

$F_{stat,der}$ derived static force [N]
$F(t)$ (measured) force on the pile head [N]
$w(t)$ (measured) displacement of the pile head [m]
$v(t)$ (measured) velocity of the pile head [m/s]
$a(t)$ (measured) acceleration of the pile head [m/s²]
t time [s]
m mass of the pile [kg]

In addition to this function, the procedure also contains a description of the method to derive the parameters.

The parametric curve $F_{stat,der}(t)$, $w(t)$ shows the derived static load-settlement curve.

In these guidelines, the use of a single degree of freedom system is applied for interpretation of the test results.

Generally, several methods for solving the axial behaviour of a pile are available. For a single pile, three methods can be adopted:

- 0-D: the motion of the pile head is considered as a single degree of freedom. The behaviour of the pile and surrounding soil are included in the description of the spring stiffness. This model is often used for the description of the derived static load-displacement curve of a pile.
- 1-D: the pile is described as a one-dimensional system. The influence of the soil around the shaft is described by springs, dashpots and sometimes masses. The soil properties (such as stiffness, strength and volumetric mass) must be translated to the properties of the springs and dashpots. This model is well-known from pile driving analysis and dynamic pile testing. For those dynamic tests, the soil properties are back-calculated by signal matching. For a rapid load test, signal matching is not required.
- 3-D: both the pile and the soil are modelled as a continuum. The properties of soil and pile (such as stiffness, strength and volumetric mass) can be directly introduced. For practical engineering the three-dimensional models are recently available. Although the mathematical description of these types of models is very accurate, the large number of parameters required and the consequences of the pile installation process on the soil behaviour, limit the accuracy that can be reached by these types of models.

It should be noted that the models described above are meant to model pile behaviour from known properties of pile and soil. However, now a so-called inverse problem is at hand: the behaviour of the pile in the soil must be derived from the measurement during an RLT.

Mostly, a rapid load test is carried out in order to answer practical questions about the static pile behaviour. These relate to the initial stiffness of the pile, when loaded by a structure, and the static bearing capacity of the pile, when loaded up to the maximum expected load. To answer these questions, knowledge of the behaviour of the pile head under mechanical load is sufficient. Unfortunately, the one- or two-dimensional model approaches discussed above have too many parameters, which cannot be derived from the pile head measurements alone.

Based on these considerations, application of a single degree of freedom system (0-D) is the most proper choice for practical engineering.

The methods described in these guidelines agree in the way the inertia effect is treated. The force due to inertia follows from the pile mass and the measured acceleration of the pile head during the test. The mass equals the pile mass. For an open ended pile, the question whether the pile is plugged during a test is relevant for calculation of the pile mass. This aspect will be discussed in Chapter 7, Section 7.1.

The methods described in these guidelines differ in the description of the rate dependency and the treatment of excess pore water pressure. The methods may be described as:

- The Unloading Point Method (UPM) considers the moment in which the velocity is zero. This model is generally best applied in sandy soils, gravely soils, silty sands and piles on rock. The soil resistance at the unloading point must still be corrected for pore water pressure effects. In order to construct the full curve between the initial loading and the pile resistance, the damping and rate effects must be taken into account.

– The Sheffield Method (SHM) is an integrated method, taking into account damping, rate effect and excess pore water pressure. This model was developed exclusively for piles installed in clay.

4.3 DYNAMIC EFFECTS

4.3.1 Inertial effects for closed piles

For very high frequency loading stress wave phenomena should be taken into account. However, the loading rate for a properly executed rapid load test is sufficient low, that the waves in the pile can be neglected. The pile and soil behave quasi-statically. This is shown theoretically by Middendorp & Daniels (1996) and empirically by Justason, Mullins et al. (1998).

Measurements by Hölscher (1995) show that wave propagation occurs in the soil. However, this appears not to have significant influence on the soil behaviour. The exact reason is currently unknown. Neglecting the wave propagation in the soil is in agreement with the approach used for the interpretation of dynamic load tests.

Although wave propagation does not appear to play a role, the inertia force from a rigid body movement of the pile must be taken into account. This force is of relatively low significance if the inertial force (from the product of mass and acceleration) is much smaller than the other forces influencing the system. A good measure of the importance of the inertial force is found from the comparison of this force with the load applied on the pile head. Neglecting the inertial force is not possible for a rapid load test.

The inertial force is calculated from the calculated pile mass and the measured acceleration. Finite element modelling of the pile driving problem ignoring additional mass contribution from the soil was shown to be valid by (Hölscher 1990), although elastic solutions show an added apparent soil mass effect due to the discretisation. Using a finer mesh reduces this numerical effect.

The force from the soil acting on the pile is calculated from the difference between the applied force on the pile head and the inertial force.

$$F_{soil}(t) = F(t) - F_{inertia}(t) = F(t) - ma(t) \qquad (4.2)$$

Where:

$F_{soil}(t)$ force from the soil to the pile [N].
$F(t)$ (measured) load on the pile head [N].
$F_{inertia}(t)$ inertial force [N].
m mass of the pile [kg].
$a(t)$ (measured) acceleration of the pile [m/s^2].

In order to obtain a reliable result, the inertia should not be dominating. A maximum value of 20% of the RLT load is reasonable.

Under the toe of a closed ended pile, a small amount of soil is also accelerated during a rapid test. This amount may be taken into account and added to the mass of the pile. A half sphere of soil with diameter equal to the equivalent diameter of the pile is an upper limit for the added mass (suggested by e.g. Matsumoto, Matsuzawa et al. (2008)). For concrete piles, the contribution is negligible, for very short light pipe piles, it might play a minor role.

4.3.2 Inertial effects for open ended pipe piles

For open ended steel pipe piles, the choice of pile mass needs additional attention. If the pile plugs, it is reasonable to include the plug mass for the interpretation of a rapid test.

The proper choice depends on the plugging of the pile. During a dynamic event, pile plugging depends on the strength of the interface between the pile and the plug and the inertia of the plug. Experience from off-shore pile driving shows that the plugging of a pile is influenced by pile length and set up time. Comparing a static load test and a rapid load test, the strength of the interface is also influenced by the rate effects in the soil.

Due to these aspects, the influence of plugging on rapid load test interpretation is uncertain. In Section 7.1 a theoretical model for plugging will be discussed.

4.3.3 Damping effects

In general terms damping is the decrease of mechanical energy from a moving or vibrating system. The mechanical energy in a system is the summation of kinetic energy and elastic energy. The mechanical energy might decrease due to transformation of mechanical energy to another form of energy within the system. This happens e.g. due to friction between soil grains, friction between water and grains, non-linear soil behaviour such as plasticity along the shaft. Another mechanism of decrease of mechanical energy from a system is related to the transport of mechanical energy through the boundary as elastic wave-radiation into the subsoil.

The damping and inertial effects together might be called the "dynamic effects" in a test. In a linear elastic system, these two effects are reasonably well defined.

4.4 RATE EFFECTS

4.4.1 General meaning of rate effect

In the literature the term rate effect is not uniquely defined, but two descriptions of this effect are meaningful:

– Rate effect is the dependency of the constitutive behaviour of a material on rate of loading.
– Rate effect is the dependency of a system on rate of loading.

The first description, introduced by (Whitman 1957), considers the rate effect as a constitutive property of a material. It means that stiffness and strength depend on loading rate. We call this the constitutive rate effect.

The second description is more general as it includes not only the constitutive property of a material, but e.g. also the damping of a system (due to wave radiation or plastic behaviour) and the pore pressure effects (considered in the next section). Normally the damping of a system is addressed separately.

In the definition of rate effects in soil, one should make a distinction between the cohesive and non-cohesive materials. For the timescale which is relevant for a rapid test, the cohesive materials, especially clay, show a strong constitutive rate effect. For non-cohesive materials the constitutive rate effect is much smaller, but for fine grained material the influence of pore water pressures might generate an additional rate effect. This effect is discussed more in detail in the next Section.

4.4.2   Measuring and modelling the rate effect

In general, two types of laboratory tests for rate effects can be distinguished:

- 'Standard' laboratory equipment (such as a direct shear box, triaxial apparatus), which are used with a high test rate. The loading rate is described as a strain rate, so the change of sample size per unit of time as a fraction of the sample size (e.g. in a triaxial test: velocity of the sample top divided by the sample height).
- Penetration equipment used in a calibration chamber, which simulates the penetration of a pile into the soil. The loading rate may be described by the pile velocity relative to the pile diameter.

Two types of relationships are used in literature: the power law relationship and the logarithmic relationship, e.g. (Charue 2004). Both relationships compare the mobilised strength with the static strength. The main differences are:

- The power law relationship assumes a constant multiplier for each doubling of the loading rate, which means an almost constant value for low loading rates.
- The logarithmic relationship assumes a constant increase for each doubling of the loading rate, which means that even for low loading rates the behaviour strongly depends on the loading rate.

However, these considerations assume a 'constant' rate effect over a large range of loading rates, which is far from reality. It must be taken into account that most relations are valid for a limited range of loading rates. Apparently, the rate effects which are observed at different loading rates might be caused by different physical mechanisms.

4.4.3   Constitutive rate effect in dry sand

For sand (Huy 2008) carried out a literature review on test results. Both types of tests (standard or penetration equipment, see section 4.4.2) are used for sand. Due to the difference in description of the loading rate, it is not known which standard tests have equal rate to comparable penetration tests. Therefore, the test results cannot be compared in absolute value. Huy also carried out laboratory tests on dry and saturated sand, in order to distinguish between rate effects and pore water effects.

In general, the constitutive rate effect in dry sand is believed to be small. Unexpected deviations are found in literature. The agreement in findings in triaxial tests is higher than for pile penetration tests.

The literature does not allow clear conclusions to be drawn on the dependence of the rate effect in sand on loading rate. Most researchers observe an increase in strength up to 5% for a factor 10 increase in loading rate relevant for RLT. At this moment this logarithmic rule seems the most appropriate description.

Where piles are installed in saturated sand the volumetric behaviour of the sand plays an important role. This will be discussed in Section 4.4.5.

4.4.4 Rate effect in clay

Rate effects in clay soils have typically been studied through the use of experimental techniques as described in Section 4.4.2. In addition, there is a significant body of previous work from field scale research studies and empirical studies from commercial projects.

From their work on high speed testing in clays and sands, Gibson & Coyle (1968) suggested that rate effects for clays could be represented by a power law relationship which Charue (2004) presents in a general form as:

$$\frac{Q_{rheol}}{Q_{stat}} = (1 + J_N \cdot \nu^N) \tag{4.3}$$

Where:

Q_{rheol} resistance of soil at an elevated velocity [N].
Q_{stat} resistance of soil at low reference velocity [N].
J_N viscous rate parameter [unit depending on N].
N power of the strain rate [–].
ν test velocity [m/s].

This work showed that rate effects were not constant for a soil as originally proposed by Smith (1962), for dynamic pile resistance but varied with rate of deformation or strain rate. This power law form of analysis was subsequently adopted and verified in several research studies notably Litkouhi & Poskitt (1980) (Model penetrometers) and Randolph & Deeks (1992)(various sources including numerical modelling) and forms the basis for the analysis technique described in Section 5.3. Typical parameters adopted for power law based analysis are shown in Charue (2004).

Alternatively, the rate effect may be represented by a logarithmic relationship which Charue (2004) represents in a generic form as:

$$\frac{Q_{rheol}}{Q_{stat}} = \left(1 + Y_L \cdot \log\left(\frac{\nu}{\nu_{ref}} \right) \right) \tag{4.4}$$

Where:

Y_L rate parameter [–].
ν_{ref} reference velocity giving the capacity Q_{stat} [m/s].
ν test velocity [m/s].

Q_{rheol} resistance of soil at an elevated velocity [N].
Q_{stat} resistance of soil at reference velocity [N].

There is a significant body of previous research in both laboratory and field studies which identify the effect of increased strain rates during loading of clay or fine grained soils, see summaries by Brown (2004) and Charue (2004). These studies have been carried out over a limited range of strain rate. Typical maximum strain rates translate to deformation rates of 0.013 mm/s which are similar to the rate experienced in "static" pile testing, but do not reflect the velocities experienced in in-situ testing (CPT) or rapid or dynamic pile testing which may be of the order of 20 to 5000 mm/s. These limited strain rate studies have led to general assumptions that the shear strength increases by 10–20% per log cycle of strain rate but this assumption only appears valid over a very narrow range of rates.

For example Dayal & Allen (1975) used an identical model to the general logarithmic form presented above (penetrometer testing, 1.3–810 mm/s) and observed a 10 to 38% increase in skin friction capacity per log cycle increase in velocity which was relatively constant at this level up to velocities between 13 and 140 mm/s. At a point within this range, the rate effect increased markedly to 93 to 100% with log cycle of velocity increase. Dayal & Allen (1975) noted that the rate parameter was affected by both the range of velocity and soil type. Skempton (1985) showed a similar jump during ring shear tests at velocities in the range of 1–2 m/s.

Although the two main forms of analysis applied to rate effects in clay differ slightly, they have common features such as the requirement for user selection of rate effect parameters. Historically, selection of such parameters for say dynamic pile testing was based upon having an appropriate database of previous testing in the material type of interest. For rapid load testing such a database of tests is not generally available and where results are available they do not contain adequate numbers of tests for clay or fine grained soils. The use of the term "clay" as a catch-all for selection of a rate parameter does not reflect the subtleties of the material where it has been shown previously that rate effects may vary significantly in the same material at different void ratios Brown & Hyde (2008).

Several previous studies have attempted to link the rate dependant parameters to easily measurable soil properties with the majority relating to Atterberg limits Brown & Hyde (2008). Gibson & Coyle (1968) and Briaud & Garland (1985) showed through laboratory element testing that rate effects for clays increased with both increasing moisture content and liquidity index and that clays of low to intermediate plasticity had significantly lower rate effects. Similar results have been shown in CPT field measurements with rate effects increasing with increasing plasticity Powell & Quaterman (1988).

Although the majority of the focus of this section has been on the effect of strain rate on ultimate capacity or behaviour, it should be noted that the rate of strain also effects soil stiffness, for example in triaxial testing of clays. Tatsuoka, Jardine et al. (1997) note at strains above 0.001% there is a 10% increase in stiffness with log cycle increase in strain rate. In current analysis, the variation of stiffness with strain rate is not considered separately which may lead to over correction for rate effects at pile working loads where accurate calculation of the equivalent static load-settlement behaviour is most important (Powell & Brown 2006). This is caused by soil rate parameters being derived from ultimate behaviour. Therefore the ideal analysis technique should also include a strain level modification of the rate effect.

4.4.5   Pore water pressures in saturated sand

The behaviour of dry granular materials (such as sand and silt) differs strongly from saturated granular materials. For practical engineering, it might be assumed that the piles are in (partly) saturated sand. This means that the volumetric behaviour of the sand plays an extremely important role. From standard geomechanics, it is known that loosely packed sand will contract during loading. This might lead to excess pore water pressures during rapid loading, and thus to a decrease of the soil strength. On the contrary, densely packed materials tend to dilate during loading, which might lead to negative excess pores pressures, and consequently to an increase of strength during rapid loading. Obviously, the behaviour of sand depends on the relative density, the loading rate and the permeability of the material.

Field measurements show the occurrence of excess pore water pressures during RLT.

Hölscher & Barends (1992) showed that during a Statnamic test on a pile in sand, excess pore water pressures and negative pore pressures are generated at a horizontal distance up to about 0.7 m from the pile toe. For the pile with an equivalent diameter of 0.28 m the consolidation time in the test was about 150 ms. From their measurements, it is concluded that in this case the soil is expected to behave in a partially drained manner.

Maeda, Muroi et al. (1998) measured the pore water pressure near to the toe of a cast-in-situ pile of 1.2 m diameter. They measured excess pressures up to 80 kPa, with a consolidation time of order 400 ms, which is longer than the duration of the test. This test can be considered as almost undrained.

Hajduk, Paikowsky et al. (1998) measured the pore water pressure in a sand layer near the pile toe. The pore water pressure was extremely small and consolidation took about 10–15 minutes. This is extremely long for a sand layer, but no explanation is given.

Matsumoto (1998) measured the pore water pressure around an open ended pile driven into soft rock. Along the pile shaft negative excess pore pressures were observed, below the pile toe positive excess pore pressures were observed. The peak values were very small with a consolidation time of approximately 200 ms.

Hölscher, Brassinga et al. (2009) describe a field test on a prefabricated concrete pile, comparable with the test of Hölscher & Barends (1992). The measurements showed similar results as in 1992. A notable detail was that the negative excess pore water pressure during unloading (at a horizontal distance of 0.7 m from the pile toe) gave such a low pressure that cavitation under the pile toe is to be expected. Almost no consolidation after the test is observed.

Laboratory experiments were conducted by Eiksund & Nordal (1996) and Hölscher, Van Lottum et al. (2008). Both studies show typical dilatancy behaviour after some elastic loading, showing that the sand around a pile tip dilates during failure. Eiksund & Nordal (1996) concludes from their 1-g tests on a 63 mm diameter pile that the generated pore water pressure has little influence on the pile bearing capacity. The measurements of Hölscher, Van Lottum et al. (2008) were analyzed by Huy (2008). He concluded that the dynamic drainage conditions play a vital role in the analysis. For a relatively large diameter pile installed in a low permeability granular soil, the loading rate influences the bearing capacity of a pile measured with a rapid test.

The dynamic drainage factor includes the influence of duration of the rapid load and the consolidation behaviour of the soil, see also Hölscher & Barends (1992). It is defined as

$$\theta = \frac{GT}{g\rho R^2} k \qquad\qquad (4.5)$$

Where:

θ dynamic drainage factor [–].
G shear modulus of the soil [Pa].
T duration of the rapid load [s].
k permeability of the soil [m/s].
g acceleration due to gravity [m/s²].
ρ volumetric mass of the water [kg/m³].
R radius of the pile [m].

The dynamic drainage factor θ is a dimensionless factor. It can be considered as the loading time relative to the consolidation time.

The description advocated by Eiksund & Nordal (1996) and van Tol, Huy et al. (2008) is valid when the soil fails and dilates during failure. This means that the influence of the pore water pressure is very limited for the stiffness, and that the installation method of the pile (soil-displacement or soil-removal) plays a role.

4.5   COMBINED EFFECTS

In soils, the dynamic effects and rate effects are all active during an RLT. It is not possible to make a unique distinction between these aspects.

A pile-soil system looses most of the mechanical energy by two mechanisms:

–   Plastic deformation in the soil around a pile.
–   Radiation of energy into the soil.

In a single degree of freedom system (SDOF-system), these two main components of damping may be represented by a dashpot (the force proportional to the velocity) and a non-linear spring constant, where permanent settlements are possible. The non-linear spring constant leads to dissipation of energy if during unloading another path is followed (non-elastic behaviour). In soil, this occurs due to a stiffer unloading behaviour. This describes elastic rebound and plastic deformation. This behaviour leads often to permanent deformation. The dashpot constant describes the energy radiation into the soil and the rate effects from the soil.

Chapter 5

Interpretation methods

5.1 INTRODUCTION

In this chapter two interpretation methods are discussed. The discussion focuses on methods with one degree of freedom, since these are generally applicable.

5.2 UNLOADING POINT METHOD

5.2.1 Definition of the method

This method was originally suggested by Middendorp, Bermingham et al. (1992). All versions of the Unloading Point Method start with determination of the so-called unloading point, where the velocity $v(t_{u\text{-}max})$ of the pile is zero. Several methods are proposed to estimate the full load-displacement curve. In these guidelines, the version which assumes a hyperbolic load-displacement curve is adopted.

This method is written as:

$$F_{stat,der}(t) = g_1[F(t), w(t), a(t); m, \eta]$$ (5.1)

Where:

$F_{stat,der}$	derived static force [N].
$F(t)$	measured force on the pile head [N].
$w(t)$	measured displacement [m].
$a(t)$	measured acceleration [m/s²].
m	mass of the pile [kg].
η	empirical parameter still depending on the soil type [-].

The empirical factor η is an empirical correction factor, which reflects the combined effect of both rate and pore pressure effects.

Two approaches are available to calculate the derived static load-displacement curve using this method. None of these methods are suitable for extrapolation to pile loads beyond the highest load tested. No method gives information about pile behaviour at loads above the derived force in the unloading point.

Method 1: predefined form of the load-displacement curve
The version of the Unloading Point Method adopted here is defined by

$$F_{derived}(t) = \frac{k_o \; u(t)}{1 + B \; u(t)}$$ (5.2)

$$B = \frac{k_0}{\eta \, [F(t_{w-max}) - m \, a(t_{w-max})]} - \frac{1}{w(t_{w-max})}$$ (5.3)

Where:

k_o initial stiffness derived from the measurement [N/m].
t_{w-max} moment the unloading point had been reached [s].

A practical application of this method has two steps:

1   At the unloading point the velocity $v(t_{w-max})$ of the pile is zero, and the derived static force at the displacement $w(t_w)$ can be estimated from

$$F_{stat,der}(t_{w-max}) = \eta \, [F(t_{w-max}) - m \, a(t_{w-max})]$$ (5.4)

2   Through this point and the calculated initial stiffness k_o a hyperbola is constructed.

If the shape of the static load-displacement curve is known, the unloading point measurement may be used to calibrate this curve on the measurement, since the determined displacement and force in the unloading point should be a point on the predetermined curve. This method is applied by Middendorp, Beck et al. (2008). They assumed a hyperbolic load-displacement curve.

The load-displacement curve for piles of the same type in similar soil conditions might be determined from available SLT. If the national code includes a method for calculation of the complete load-displacement diagram, the unloading point measurement may be used to calibrate this diagram on the result of the RLT (see Appendix D).

Method 2: a multi cycle test
The rapid load test is carried out with several loading cycles with increasing peak loads. For each loading cycle the unloading point and the mobilised static resistance in the soil are determined.

For each cycle, the unloading point (with velocity $v(t_{w-max})$ of the pile is zero), and the derived static force at the displacement $w(t_w)$ can be estimated from

$$F_{stat,der}(t_{w-max}) = \eta \, [F(t_{w-max}) - m \, a(t_{w-max})]$$ (5.5)

The derived static load-displacement curve can be constructed by a best fit curve through the derived points or by a best fit hyperbolic curve as described in method 1.

The multiple cycle results must be considered as reloading events on the pile, so the cumulative displacement over the loading cycles must be considered (Schmuker 2005). Therefore, all steps of the test must be carried out with an increasing peak load.

5.2.2   Measurement/estimation of required parameters

The empirical factor η depends on soil type. Its value can be estimated from field tests on sites, where both the static and rapid tests are carried out. This approach was adopted by McVay, Kuo et al. (2003) and Hölscher and van Tol (2009b). Appendix C shows the background. The empirical factor depends on type of pile installation and the type of soil.

The following empirical factors η are suggested:

The number of sites has to be used if a characteristic value is calculated using the Student-t distribution. A log normal distribution is used, since it relates to a factor with a relatively large coefficient of variation.

Table 5.1 shows a large coefficient of variation for piles in clay. In Section 5.3 a more appropriate method for clay soils is described. Appendix H shows a discussion on the empirical factor for piles in clay, which is based on Section 5.3. The procedure of Appendix H is not tested, but this procedure is expected to give less variation.

Table 5.1 Summary.

Pile type Soil type	All Clay	Displacement Sand	Bored and CFA Sand
Empirical factor η	0.66	0.94	1.06
Standard deviation	0.32	0.15	0.28
Coefficient of variation	0.49	0.15	0.27
Number of cases	12	21	11
Number of sites	6	10	4

5.2.3   Example of the application

An example of the UPM method is given in Appendix A.

5.3   SHEFFIELD METHOD

5.3.1   Definition of the method

The method developed at Sheffield University by Brown (2004) and Brown, Hyde et al. (2006) is written as:

$$F_{stat,der}(t) = g_2[F(t), v(t), a(t); m, \alpha, \beta, v_{static}, v_{ref}] \tag{5.6}$$

Where:

$F_{stat,der}$ derived static force [N].
$v(t)$ velocity of the pile [m/s].
$a(t)$ acceleration of the pile [m/s^2].
m mass of the pile [kg].
α and β soil dependent model parameters [-].
v_{static} loading rate, which is defined as static [$1*10^{-5}$ m/s].
v_{ref} normalizing value [1 m/s].

The modified version (Brown & Hyde, 2008) incorporates the variation of the rate effect parameters with pile settlement or soil strain level. Balderas-Meca (2004) suggested linear variation of α up to approximately 1.0 to 1.2% of the pile settlement relative to the pile diameter. Above this level, consistent with ultimate pile behaviour, the value of α becomes constant. To incorporate this variation in α, the following equation can be used:

$$F_{stat,der}(t) = \frac{F(t) - m\,a(t)}{1 + \dfrac{F(t)}{F_{max}} \alpha \left[\left(\dfrac{\tilde{v}(t)}{v_{ref}} \right)^{\beta} - \left(\dfrac{v_{static}}{v_{ref}} \right)^{\beta} \right]} \qquad (5.7)$$

Where:

F_{max} maximum rapid load applied during testing [N].
$\tilde{v}(t)$ modified velocity function, defined by

$$\tilde{v}(t) = |v(t)| \quad \text{if} \quad t < t_{max}$$
$$\tilde{v}(t) = v_{max} \quad \text{if} \quad t \geq t_{max}$$

if $\tilde{v}(t) < v_{static}$, v_{static} must be used to avoid corrections values smaller than 1.

Where:

v_{max} maximum value of the downward velocity of the pile before the unloading point has been reached [m/s].
t_{max} time at which v_{max} is reached [s].

Here it is assumed that at peak rapid load (F_{max}) the pile has been significantly mobilised and that the value of α has become constant.

This method of analysis is only valid for piles installed in clay where the majority of their capacity is derived from the shaft resistance. The method has not been validated for other soil types. If this approach is used on a pile with significant tip capacity, it is likely to underestimate the static ultimate capacity as rate effects are generally significantly reduced for pile tip resistance. The method is valid for the branch of the rapid load displacement until the unloading point (so downwards movement only).

The tests which can be used for the determination of the model parameters α and β are discussed in detail in Appendix E.

5.3.2   Measurement or estimation of required parameters

The performance of the analysis depends on the selection of appropriate soil specific rate parameters. The parameters available for selection are mainly based upon laboratory testing of clays using high speed triaxial and penetrometer testing and model scale pile testing (Brown 2004), although there are limited examples from field scale studies. In all cases the parameters are defined for velocity ranges appropriate to rapid load testing. The parameters are summarized in the following table.

Table 5.2 Summary of rate parameters from previous studies (Powell & Brown 2006).

Originator	Soil	Index properties (LL, PL, PI-%)	α [-]	β[-]	Test conditions
Randolph &	Sand	–	0.1	0.2	Summary of previous
Deeks	Clay	–	1.0	0.2	work
Balderas-Meca	Grimsby glacial till	20–36, 12–18, 7–20	0.9	0.2	Statnamic tests
Brown	Model clay	37, 17, 20	1.26	0.34	Model Statnamic tests
Poskitt & Leonard	Cowden till	40, 20, 20	1.0	0.27	Penetrometer tests
Litkouhi & Poskitt	London clay	70, 27, 43	1.77	0.18	Penetrometer tests
	Forties clay	30, 20, 18	0.99	0.23	
	Magnus clay	31, 17, 14	0.86	0.46	

The parameters themselves have been obtained in one or two ways. For the majority of the laboratory studies representation of rate behaviour has been determined from multiple tests in a fixed soil type with each test in that soil type being at a constant strain rate. These tests are then repeated at different constant strain rates to build up a rate effect model that covers the spectrum of strain rates of interest. This process is then repeated in different soil types. The method that is generally employed for field data is back analysis where a direct comparison is made between a rapid load test and a static pile test.

In order to aid selection of parameters based upon the summary of tests above the rate parameter β was set at 0.2 and α made soil specific by relating it to the soils plasticity index which can be written as:

$$\alpha = 0.031PI + 0.46 \tag{5.8}$$

Where:

PI   plasticity index [%].

This relationship is tentative as it is based upon a limited data set (Powell & Brown 2006). The relationship covers plasticity index ranging from 14 to 43% but the majority of the data set is for soils with PI in the range 14–20%.

However it should be noted that:

– The majority of studies that make comparisons between static and rapid soil behaviour do so using strain controlled testing i.e. constant strain rate. This implies that the results of the above analysis should be more comparable with constant rate of penetration type static pile tests rather than maintained load tests (load control).
– The value α depends on both the parameters v_{static} and v_{ref} used for the interpretation of the test results. v_{ref} cannot be freely chosen.

Chapter 6

Safety approach

6.1 METHOD

The recommended safety approach is based on the regulations of Eurocode 7.1. In Appendix A of this code, the partial resistance factors and the correlation factors to derive characteristic and design values are presented for the Static Load Test (SLT) and the Dynamic Load Test (DLT). No factors related to RLT are presented. Because the character of RLT is between SLT and DLT, the recommended safety approach is also in between SLT and DLT.

6.2 PARTIAL FACTORS

We assumed that N RLT's are carried out. Each test is interpreted according to the guidelines in Chapter 5. These lead to N values for $R_{c;RLT;i}$

Where:

$R_{c;RLT;i}$ derived static ultimate capacity of Rapid Load Test I [N].
i   test number (1 to N).

The following method is recommended:

1   *Determine characteristic value as the minimum from:*

$$R_{c;k} = Min \{R_{c;rlt;avg}/\xi_1, R_{c;rlt;min}/\xi_2\}$$
(6.1)

Where:

N   number of tested piles [-].
$R_{c;RLT;avg}$ mean value of derived ultimate static capacity of N tests [N].
$R_{c;RLT;min}$ minimum value of derived ultimate static capacity of N tests [N].
$R_{c;k}$ characteristic value of ultimate pile capacity [N].
ξ_1 and ξ_2 correlation factors, to be applied on respectively $R_{c;RLT;avg}$ and $R_{c;RLT;min}$ (statistical factors to assess characteristic value of capacity from tests) [-].

The correlation factors $\xi_{1;RLT}$ and $\xi_{2;RLT}$ are calculated from the recommended values for static and dynamic load testing by

$$\xi_{1,RLT} = \frac{1}{2}(\xi_{1;SLT} + 0.85 \cdot \xi_{5;DLT}) \tag{6.2}$$

$$\xi_{2,RLT} = \frac{1}{2}(\xi_{2;SLT} + 0.85 \cdot \xi_{6;DLT}) \tag{6.3}$$

Where:

$\xi_{i,SLT}$ and $\xi_{i,DLT}$ recommended values in Eurocode 7.1 (Annex A, Table A9 and A11, or to be determined by member states in the National Annex).

Remarks about DLT:

– An empirical factor of 0.85 for DLT is only applicable when signal matching is used by measuring the extension and acceleration during the test.
– In table A11 of Annex A of Eurocode 7.1 only correlation factors are presented for 2 or more tests. Therefore, it is recommended to perform at least 2 RLT tests.

Figure 6.1 The rapid load test can be applied for checking the initial stiffness and bearing capacity of piles.
Copyright: Allnamics Pile Testing Experts BV and Profound BV.

2   *Calculate the design value following from:*

$$R_{c;d} = \frac{R_{c;k}}{\gamma_R}$$
(6.4)

Where:

γ_R partial safety factor. Recommended values in Eurocode 7.1 (Annex A, Tables A6, A7 and A8, or to be determined by member countries in the National Annex).

An example of the application is given in Appendix F.

Chapter 7

Special aspects

7.1 OPEN-ENDED PILES

The essential difference between a closed-ended pile and open ended-pile is the question whether or not the pile plugs during installation, testing or functioning. The question arises whether the pile plugs during an RLT in a similar way as during an SLT.

Two aspects are of interest:

– The mass effect of the plug
– The strength of the plug-pile interface

Ochiai, Kusakabe et al. (1997) studied this aspect, both experimentally and numerically. An open ended steel pile (length 47.6 m, diameter 1.5 m and wall thickness 22 mm) was driven in diluvial sand and clay layers with cone resistance $q_c = 5$–12 MPa. The pile head was about 12 m above the water table, the water depth was about 8 m, so the first 20 m of the pile had no shaft friction at all. The pile was tested by Statnamic one week after driving and dynamically one week after the Statnamic test. Ochiai, Kusakabe et al. (1997) described the system by two coupled one-dimensional mass-spring systems. The first system describes the pile. This system is identical to the one used for closed-ended piles, the outer shaft resistance can be described as usual. The second mass-spring system describes the behaviour of the plug. This seems a reasonable model.

During the dynamic test the pile does not plug, while during the Statnamic test the pile plugs. Plugging is expected during Static Test. The model describes the test results well.

Another open ended steel pile (length 11.0 m, diameter 0.4 m, and wall thickness 12 mm) was installed in a clayey soil with $q_c = 3$ MPa (Matsumoto, Michi et al., 1995, Matsumoto, Muroi et al., 1998). This pile was tested statically after driving and by Statnamic 14 months later. Both tests showed plugging. In this case, the soil behaved undrained during the static as well as the Statnamic test. In order to derive the static bearing capacity of the test, the influence of excess pore water pressure must be taken into account.

In these guidelines, the interpretation of an RLT is undertaken by a single degree of freedom system. Matsumoto's model suggests the application of a two degrees of freedom system. However, the wave speed in the soil plug is much lower than the wave speed in the pile. Therefore, wave phenomena might start to play a role in the soil plug, since in most off-shore applications the piles are relatively long. Wave effects in the plug may influence the vertical stress in the plug. Therefore this effect is expected to have more influence on a sand plug than on a clay plug. If, due to the stress-wave the normal effective stress decreases, the strength of the sand decreases, which leads to an underestimation of the influence of the plug.

The rate effects in soil are discussed in Section 3.2.2. Important differences between sand and clay are observed.

For sand, the pore water pressure plays a role. Due to the length of the pile, the plug will behave in an undrained manner during an RLT. During an SLT, the plug will behave partly or even fully drained. Due to the undrained behaviour, the strength of the sand increases, so the pile might plug during an RLT, whereas that will not be the case during an SLT. The RLT may overestimate the capacity.

For clay, the rate effect leads to an increase of the strength of the clay. Therefore, also in clay, the RLT might lead to an overestimation of the capacity of a pile.

At the moment, the application of RLT to open-ended piles is still not fully understood. More numerical and empirical research is needed. Application of RLT to open-ended piles must be treated with additional care.

7.2 INFLUENCE OF PILE INSTALLATION

Several pile types are used in practice. A general distinction is prefabricated displacement piles, in-situ fabricated displacement piles and non-displacement piles (bored and CFA).

Appendix B shows the differences for soil displacement piles and piles with soil removal for sand. The differences might, moreover, be dependant on soil type.

The centrifuge tests reported by Huy (2008) show that a prefabricated pile installed in loose sand, leads to dilating behaviour during loading. This is contrary to expectations. A non-displacement pile might lead to the opposite behaviour. Further work is required to verify these findings.

Heerema (1979) shows that the rate effect in clay is dependent on stress level. Since pile installation strongly influences the stress level in the surrounding soil close to the pile, it is expected that also installation effects play a role in clay. However, this is not discussed in literature.

To be on the safe side, it is advised to use correction factors for both interpretation methods only after calibration with the same type of installation as the piles under consideration.

7.3 LAYERED SOILS

The interpretation methods were originally designed for the analysis of piles in homogeneous soil. However, in practical engineering the soil is often a layered structure. The following adjustments for the methods are suggested.

UPM Method

From a static calculation the distribution of the soil resistance under characteristic load is estimated, together with the contributions of the shaft and toe. The correction factor can be estimated from:

$$\eta = \frac{F_{stat,clay}}{F_{stat,sum}}\eta_{clay} + \frac{F_{stat,sand}}{F_{stat,sum}}\eta_{sand}$$

(7.1)

Where:

$$F_{stat,sum} = F_{stat,clay} + F_{stat,sand}$$

$F_{stat,clay}$ is the calculated static resistance from the clay layer(s) in the soil profile [N]
$F_{stat,sand}$ is the calculated static resistance from the sand layer(s) in the soil profile [N]

This method is also suggested and evaluated by Weaver and Rollins (2010).

Sheffield Method

If the clay along the shaft has essentially different parameters (α and β) a similar calculation method can be used for the correction factor in the enumerator of the equation

$$F_{stat;der} = \frac{F_{RLT}(t) - ma(t)}{1 + \dfrac{F(t)}{F_{max}}\displaystyle\sum_{j=1}^{N}\frac{F_{stat;j}}{F_{stat;sum}}\alpha_j\left[\left(\dfrac{\tilde{v}(t)}{v_{ref}}\right)^{\beta_j} - \left(\dfrac{v_{static}}{v_{ref}}\right)^{\beta_j}\right]}$$

(7.2)

Where:

$F_{stat;der}$ derived static force [N].
$F_{RLT}(t)$ force on the pile top [N].
F_{max} maximum rapid load applied during testing [N].
$a(t)$ acceleration of the pile [m/s²].
m mass of the pile [kg].
N the number of layers along the shaft [-]
α_j and β_j soil depending model parameters for layer j [-].
v_{static} value of the loading speed, which is defined as static ($1*10^{-5}$ m/s).
v_{ref} normalizing value (1 m/s).
$F_{stat;j}$ calculated resistance from layer j [N].
$F_{stat;sum}$ the summation of the calculated resistance over all layers [N].

The variable $\tilde{v}(t)$ is explained under equation 5.7, page 24.
The parameters must be known for each layer j.

These equations are derived from engineering judgment, the accuracy must not be overestimated and the results are not yet tested against measurements.

7.4   OTHER ASPECTS

The unloading cycle of a RLT may give insight in the initial stiffness of the pile after loading. This aspect is completely ignored in the interpretation methods presented in these guidelines. The initial part of the unloading (directly after passing the unloading point) must be handled with care, since the physics in that part are not fully understood.

The full loading cycle should be inspected visually, in order to avoid misinterpretation due to pile defects. Sometimes, pile defects are visible from deviations in the unloading branch of the test.

Chapter 8

Symbols and definitions

8.1 SYMBOLS

a	pile acceleration [m/s^2].
F	force [N].
F_c	characteristic load on pile [N].
$F_{inertia}$	inertia force (equals pile mass m times acceleration a) [N].
$F_{damping}$	damping force depending on the pile velocity [N].
F_{max}	maximum rapid load applied during testing [N].
F_{RLT}	force on the pile top by an RLT device, measured [N].
F_{soil}	soil forces acting on the pile [N].
$F_{stat,clay}$	calculated static resistance form the clay layer(s) in the soil profile [N].
$F_{stat,der}$	equivalent static force derived from the results of a rapid load test [N].
F_{static}	force in a static test [N].
$F_{stat,sand}$	calculated static resistance from the sand layer(s) in the soil profile [N].
$F_{stat,sum}$	sum of the calculated static resistance in all layer(s) in the soil profile [N].
G	shear modulus of the soil [Pa].
g	acceleration due to gravity [m/s^2].
J_N	viscous rate parameter [unit depending on N].
k	permeability of the soil [m/s].
k_c	initial stiffness, stiffness up to the characteristic load [N/m].
k_o	initial stiffness derived from the measurement [N/m].
m	pile mass [kg].
N	power of the strain rate [–].
N	number of tested piles [–].
N	number of layers [–].
Q_{rheol}	resistance of soil at an elevated velocity [N].
Q_{stat}	resistance of soil at low reference velocity [N].
R	radius of the pile [m].
$R_{c;RLT;avg}$	mean value of derived ultimate static capacity of N tests [N].
$R_{c;RLT;min}$	minimum value of derived ultimate static capacity of N tests [N].
$R_{c;k}$	characteristic value of ultimate pile capacity [N].
T	duration of the rapid load [s].
Y_L	rate parameter for logarithmic relationship [–].
t	time [s].
t_{u-max}	moment the unloading point had been reached [s].

v_{ref}	normalizing value for velocity [1 m/s].
v	pile velocity [m/s].
v_{static}	loading speed defined as static [$1*10^{-5}$ m/s].
$\tilde{v}$	modified velocity function [m/s].
w_c	displacement at characteristic load [m].
w	pile displacement [m].
α	material dependant model parameter [–].
β	material dependant model parameter [–].
ξ_1	correlation factors, to be applied on $R_{c;RLT;avg}$ [–].
ξ_2	correlation factor, to be applied on $R_{c;RLT;min}$ [–].
$\xi_{i,SLT}$	recommended value in Eurocode 7.1 [–].
$\xi_{i,DLT}$	recommended value in Eurocode 7.1 [–].
η	empirical parameter which depends on the soil type [–].
ρ	volumetric mass of the water [kg/m³].
θ	dynamic drainage factor [–].

8.2 DEFINITIONS

Rapid Load Test
 a test that fulfils the requirements in the European Standard (the draft is under discussion in Working group 8 of TC341 of CEN. The document prepared by the CUR committee H410 is presented in the appendix of Hölscher and Tol (2009) See also Appendix I.

Static Load Test
 a test that fulfills the requirements in the National Standard for Static Load Testing on piles under compression (see Appendix I).

Static load-settlement diagram
 the load settlement behaviour of a pile in the soil measured by a Static load test conforming to National Standard for Static Load Testing on piles under compression (see Appendix I).

Rapid load-settlement diagram
 the diagram of the force on the pile head and the settlement of the pile head measured during a Rapid Load Test.

Derived load-settlement diagram
 the approximated Static load-settlement diagram estimated from the rapid load-settlement diagram.

Single cycle method
 an interpretation method which calculates the derived load-settlement diagram from one blow on the test pile.

Multi cycle method
    an interpretation method which calculates the derived load-settlement diagram from a number of blows on the test pile.

Rate effect
    the dependency of soil mechanical behaviour (strength and/or stiffness) on loading rate.

Dynamic effects
    the summed effect of damping and inertia.

Damping
    the loss of mechanical energy from a vibrating system. For a free vibrating system, this loss might be due to transformation of mechanical energy to another type of energy (e.g. heat), this is known as material damping, or due to transport of mechanical energy through the boundary of the system, this is known as radiation damping.

Inertia
    the effect that a finite force on a body leads to a finite acceleration of the body.

Inertial force
    the force which is derived from the mass multiplied by the acceleration.

Constitutive rate effect
    the part of the rate effect which is observed in a material which is independent of the size of the sample.

Pore water pressure effect
    the effect that the mechanical behaviour (strength and/or stiffness) of a soil is influenced by the generation of excess pore water pressure.

Unloading point
    moment in an RLT where the velocity of the pile is zero, this corresponds with the time of maximum displacement.

References and Literature

Balderas-Meca, J. (2004) Rate effects in rapid loading of clay soils. PhD-thesis University of Sheffield. Sheffield, UK.

Briaud, J. L. & Garland, E. (1985) Loading rate method for pile response in clay. *ASCE Journal of Geotechnical Engineering*, 111 (3), 319–335.

Briaud, J. L. & Ballouz, M., et al. (2000) Static capacity predictions by dynamic methods for three bored piles'. *Journal of Geotechnical and Geoenvironmental engineering* 126 (7), 640–649.

Brown, M. J. (2004) The rapid load testing of piles in fine grained soils. PhD-thesis University of Sheffield, Sheffield, UK.

Brown, M. J. & Hyde, A. F. L. (2008) Rate effects from pile shaft resistance measurements. *Canadian Geotechnical Journal*, 45 (3), 425–431.

Brown, M. J. & Hyde, A. F. L., et al. (2006) Instrumented rapid load pile tests in clay. *Geotechnique*, 56 (9), 627–638.

Charue, N. (2004) Loading rate effects on pile load-displacement behaviour derived from back-analysis of two load testing procedures, PhD-thesis Catholique University of Louvain. Louvain, BE.

Dayal, U. & Allen, J. H. (1975) The effect of penetration rate on the strength of remoulded clay and sand samples. *Canadian Geotechnical Journal*, 12, 336–348.

Eiksund, G. & Nordal, S. (1996) Dynamic model pile testing with pore pressure measurements. *5th Int Conf. application of stress wave theory to piles*. Florida, USA.

Gibson, G. C. & Coyle, H. M. (1968) *Soil damping constants related to common soil properties in sands and clays (Bearing capacity for axially loaded piles)*. Texas Transportation Institute Research Report 125-1 (Study 2-5-67-125). Texas, Texas A&M University.

Hajduk, E. L. & Paikowsky, S. G., et al. (1998) The behaviour of piles in clay during Statnamic, dynamic and different static load testing procedures. *2nd int. Statnamic seminar*. Tokyo, Japan.

Heerema, E. P. (1979) Relationships between wallfriction, displacement, velocity and horizontal stress in clay and sand for pile driveability analysis. *Ground Engineering* 12 (1), 55–61.

Heijnen, W. J. & Joustra, K. (eds.) (1996) *Application of stress wave theory to piles: Test results*, Balkema, Rotterdam, NL.

Holeyman, A. & Couvreur, J. M., et al. (2001) Results of dynamic and kinetic pile load tests and outcome of an International prediction event; Screw piles, Technology, installation and design in stiff clay In: Holeyman, A. (ed.), Balkema, Lisse, NL.

Holeyman, A. & Charue, N. (2003) International pile capacity prediction event at Limelette, Belgian Screw pile Technology, design and recent developments, In: Maertens & Huybrechts (eds.), Balkema, Lisse, NL.

Horikoshi, K. & Kato, K., et al. (1998) Finite element analysis of Statnamic loading test of pile. *2nd Int. Statnamic seminar*, Tokyo, Japan. 295–302.

Hölscher, P. (1995) Dynamical response of saturated and dry soils. PhD-thesis Delft University of Technology. PhD178, Delft, NL.

Hölscher, P. & Barends, F. B. J. (1992) Statnamic load testing of foundation piles. *Fourth Intern. Conf. on application of stress wave theory to piles.* The Hague, the Netherlands, Balkema.

Hölscher, P. & Van Lottum, H., et al. (2008) Rapid pile test simulation in the GeoCentrifuge. *2nd BGA International Conference on Foundations.* Dundee, Scotland.

Hölscher, P. & Van Tol, A. F. (2009a) Recent advances in rapid load testing on piles. Hölscher, P. & Van Tol, A. F. Francis Taylor.

Hölscher, P. & Van Tol, A. F. (2009b) Database of Field measurements of SLT and RLT for calibration In: Hölscher, P., Van Tol, A. F. (eds.), *Rapid load testing on piles* Taylor and Francis.

Hölscher, P. & Brassinga, H. E., et al. (2009) Rapid load field tests interpreted with the new Guideline. *Proceedings of the 17th International Conference on Soil Mechanics and Geotechnical Engineering, 5–9 October 1275–1278,* Alexandria, Egypt.

Horvath, J. S. & Trochalides, T., et al. (2004) Axial compressive capacities of a new tapered steel pipe pile at the John F. Kennedy international airport. *Proc. 5th Int. Conf. on case histories in geotechnical engineering,* New York, NY.

Huy, N. Q. (2008) Rapid load testing of piles in sand; Effects of loading rate and excess pore pressure. PhD-thesis Delft University of Technology. Delft, NL.

Janes, M. C. & Bermingham, P.D., et al. (1991) An innovative dynamic test method for piles. *Proc. 2nd Int. Conf. Recent advances in geotechnical earthquake engineering and soil dynamics.* St. Louis, Missouri, USA.

Justason, M. D. & Mullins, G., et al. (1998) A comparison of static and Statnamic load tests in sand: A case study of the Bayou Chico bridge in Pensacola, Florida. Statnamic loading test '98, Balkema, Rotterdam.

Litkouhi, S. & Poskitt, T. J. (1980) Damping constants for pile driveability calculations, *Geotechnique,* 30 (1), 77–86.

Maeda, Y. & Muroi, T., et al. (1998) Applicability of Unloading-Point-Method and signal matching analysis on Statnamic test for cast-in-place pile. *2nd Int. Statnamic seminar.* Tokyo, Japan. pp. 99–107.

Matsumoto, T. & Michi, Y., et al. (1995) Performance of axially loaded steel pipe piles driven in soft rock. *J. Geotech. Eng.,* 121 (4), 305–315.

Matsumoto, T. (1998) A FEM analysis of a Statnamic test on open-ended steel pipe pile. *2nd Int. Statnamic seminar,* Tokyo, Japan, 287–294.

Matsumoto, T. & Matsuzawa, K., et al. (2008) A role of pile load test – Pile load test as element test for design of foundation system. *Proc 8th Int. Conf. on The Application of Stress-Wave Theory to Piles: Science, Technology and Practice.* Lisbon, Portugal.

McVay, M. & Kuo, C. L., et al. (2003) Calibrating resistance factors for load and resistance factor design for statnamic load testing report University of Florida.

Middendorp, P. (2005) Verwendung von statnamischen Probebelastungen in Deutschland, Pfahl-Symposium 2005, Braunschweig 2005.

Middendorp, P. & Hölscher, P., et al. (2009) Erfahrungen und Entwicklungen mit stat-namischen Probebelastungen, Pfahl-Symposium 2009, Braunschweig BRD (in German).

Middendorp, P. & Beck, C., et al. (2008) Verification of Statnamic load testing with static load testing in a cohesive soil type in Germany. *Proc. Int. Conf. on the application of stress-wave theory to piles.* Lisbon, Portugal, Millpress Science Publisher.

Middendorp, P. (2010) Re-analysis of the Rapid Load Testing results Waddinxveen, 1996, report Profound R07DM010.

Middendorp, P. & Bermingham P., et al. (1992) Statnamic load testing of foundation piles. Application of stress-wave theory to piles. The Hague, the Netherlands, Balkema, Rotterdam, the Netherlands. pp. 581–588.

Middendorp, P. & Daniels, B. (1996) The influence of stress wave phenomena during Statnamic load testing. *Fifth International conference on the application of stress-wave theory to piles*. Orlando, Florida, USA.

Ochiai, H. & Kusakabe, O., et al. (1997) Dynamic and Statnamic load tests on offhore steel pipe piles with regard on failure mechanisms of pile-soil interfaces at external and internal shafts, *Int. Conf. on Foundation Failures*, May 12–13, pp. 327–338.

Opstal, A. Th. P. J. & Van dalen, J. H. (1996) proefbelasting op de maasvlakte (load tests in the maasvlakte (= new rotterdam harbour area)), cement, (2), 53–62. in Dutch

Poskitt, T. J. & Leonard, C. (1982) Effect of velocity on penetrometer resistance. *Proc. of the European Symposium on penetration testing*, ESOPT-2, Amsterdam, NL. pp. 331–336.

Powell, J. J. M. & Quaterman, R. S. T. (1988) The interpretation of cone penetration in clays with particular reference to rate effects. *Proc. of the Int. Symp. on penetration testing*, Orlando, Florida, ISOPT-1, A.A. Balkema, Rotterdam, pp. 903–910.

Powell, J. J. M. & Brown, M. J. (2006) Statnamic pile testing for foundation re-use. In: Butcher, A.P., Powell, J. J. M. & Skinner, H. D., (eds.) *Int. Conf. on the re-use of foundations for urban sites*, Watford, UK. pp. 223–236.

Randolph, M. F. & Deeks, A. J. (1992) Dynamic and static soil models for axial pile response. *Proc. 4th Int. Conf. Appl. Stress-Wave Theory to Piles*, The Hague, Rotterdam, Balkema. pp. 3–14.

Romano, M. C. & Middendorp, P. (2010) Pile Foundation design philosophy and testing program for a new generation diesel fuel plant, *DFI Conference, Geotechnical Challenges in Urban Regeneration*, London, UK.

Schellingerhout, A. J. G. & Revoort, E. (1996) Pseudo static pile load tester. *Proc. 5th Int. Conf. Application of stress-wave theory to piles*. Florida, USA.

Schmuker, C. (2005) Vergleich statischer und Statnamischer Pfahlprobebelastung. MSc-thesis bauingenieurwesen. Biberach, Germany.

Skempton, A. W. (1985) Residual strength of clays in landslides, folded strata and the laboratory. *Geotechnique*, 35 (1), March, 3–18.

Standard, (2008) Geotechnical investigation and Testing of geotechnical structures Testing of piles: rapid load testing draft. Deltares.

Smith, E. A. L. (1962) Pile-driving analysis by the wave equation. *ASCE, Journal of Soil Mechanics and Foundation division*, 127 (I), (Paper 3306), 1145.

Tatsuoka, F. & Jardine, R. J. et al. (1997) Characterising the pre-failure deformation properties of geomaterials. *Proc. 14th Int. Conf. on Soil Mechanics and Foundation Engineering*, Hamburg, 2129–2164.

Tol, A. F. & van, N. Q. Huy et al. (2009) Rapid model pile load tests in the geotechnical centrifuge (2): Pore pressure distribution and effects. In: Hölscher, P. & Tol, A. F. van (eds.), Rapid load testing on piles Taylor and Francis.

Weaver, T. J. & Rollins, K. M. (2010) Reduction Factor for the Unloading Point Method at Clay Soil Sites, *Journal of Geotechnical and Environmental Engineering, ASCE*, April 2010, 643–646.

Whitman, R. V. (1957) The behaviour of soils under transient loadings. *Proc. 4th Int. Conf. on Soil Mechanics and Foundation Engineering*, pp. 207–210.

Appendix A

Step by step description of the Unloading Point Method (UPM)

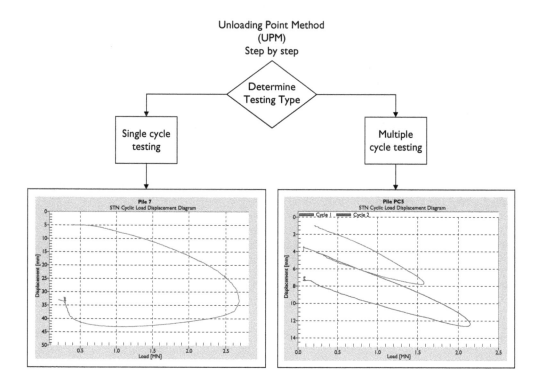

Single cycle testing (UPM)
Overview all steps

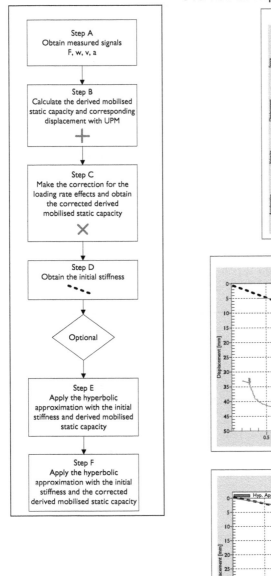

Step A
Obtain measured signals
F, w, v, a

Step B
Calculate the derived mobilised static capacity and corresponding displacement with UPM

$+$

Step C
Make the correction for the loading rate effects and obtain the corrected derived mobilised static capacity

$\times$

Step D
Obtain the initial stiffness

Optional

Step E
Apply the hyperbolic approximation with the initial stiffness and derived mobilised static capacity

Step F
Apply the hyperbolic approximation with the initial stiffness and the corrected derived mobilised static capacity

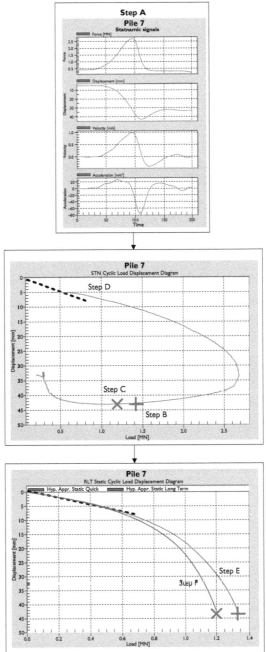

Single cycle testing (UPM),
Steps B and C.
Mobilised static capacity

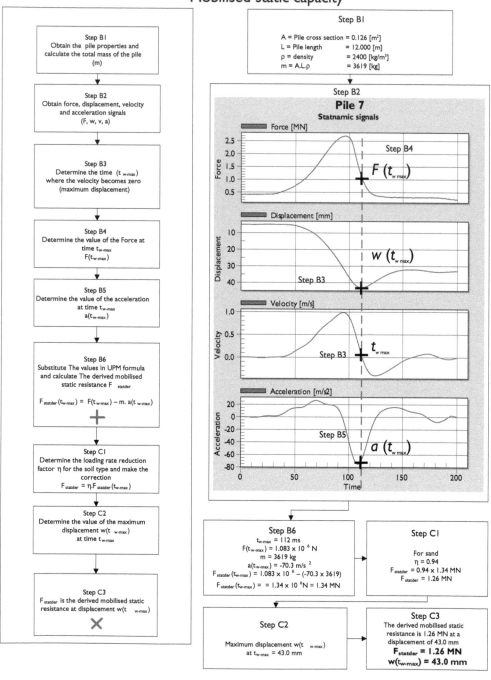

Step B1
Obtain the pile properties and calculate the total mass of the pile (m)

Step B2
Obtain force, displacement, velocity and acceleration signals
(F, w, v, a)

Step B3
Determine the time (t_{w-max}) where the velocity becomes zero (maximum displacement)

Step B4
Determine the value of the Force at time t_{w-max}
$F(t_{w-max})$

Step B5
Determine the value of the acceleration at time t_{w-max}
$a(t_{w-max})$

Step B6
Substitute The values in UPM formula and calculate The derived mobilised static resistance $F_{statder}$

$F_{statder}(t_{w-max}) = F(t_{w-max}) - m. a(t_{w-max})$

Step C1
Determine the loading rate reduction factor η for the soil type and make the correction
$F_{statder} = \eta.F_{statder}(t_{w-max})$

Step C2
Determine the value of the maximum displacement $w(t_{w-max})$ at time t_{w-max}

Step C3
$F_{statder}$ is the derived mobilised static resistance at displacement $w(t_{w-max})$

Step B1

A = Pile cross section = 0.126 [m²]
L = Pile length = 12.000 [m]
ρ = density = 2400 [kg/m³]
m = A.L.ρ = 3619 [kg]

Step B2

Pile 7
Statnamic signals

Step B6
t_{w-max} = 112 ms
$F(t_{w-max})$ = 1.083 x 10^6 N
m = 3619 kg
$a(t_{w-max})$ = -70.3 m/s²
$F_{statder}(t_{w-max})$ = 1.083 x 10^6 − (−70.3 x 3619)
$F_{statder}(t_{w-max})$ = = 1.34 x 10^{6}N = 1.34 MN

Step C1

For sand
η = 0.94
$F_{statder}$ = 0.94 x 1.34 MN
$F_{statder}$ = 1.26 MN

Step C2

Maximum displacement $w(t_{w-max})$
at t_{w-max} = 43.0 mm

Step C3
The derived mobilised static resistance is 1.26 MN at a displacement of 43.0 mm
$F_{statder}$ = **1.26 MN**
$w(t_{w-max})$ = **43.0 mm**

Single cycle testing (UPM)
Step D
Initial stiffness

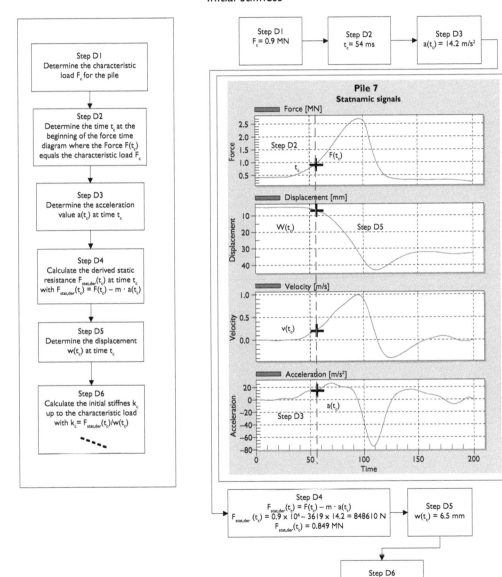

Step D1
Determine the characteristic load F_c for the pile

Step D2
Determine the time t_c at the beginning of the force time diagram where the Force $F(t_c)$ equals the characteristic load F_c

Step D3
Determine the acceleration value $a(t_c)$ at time t_c

Step D4
Calculate the derived static resistance $F_{stat,der}(t_c)$ at time t_c with $F_{stat,der}(t_c) = F(t_c) - m \cdot a(t_c)$

Step D5
Determine the displacement $w(t_c)$ at time t_c

Step D6
Calculate the initial stiffnes k_c up to the characteristic load with $k_c = F_{stat,der}(t_c)/w(t_c)$

Step D1
$F_c = 0.9$ MN

Step D2
$t_c = 54$ ms

Step D3
$a(t_c) = 14.2$ m/s²

Pile 7
Statnamic signals

Force [MN]

Step D2

$F(t_c)$

t_c

Displacement [mm]

$W(t_c)$

Step D5

Velocity [m/s]

$v(t_c)$

Acceleration [m/s²]

$a(t_c)$

Step D3

Time

Step D4
$F_{stat,der}(t_c) = F(t_c) - m \cdot a(t_c)$
$F_{stat,der}(t_c) = 0.9 \times 10^6 - 3619 \times 14.2 = 848610$ N
$F_{stat,der}(t_c) = 0.849$ MN

Step D5
$w(t_c) = 6.5$ mm

Step D6
$k_c = 0.849/6.5$
$k_c = 0.131$ **MN/mm**

Single cycle testing (UPM)
Steps E and F
Hyperbolic Approximation

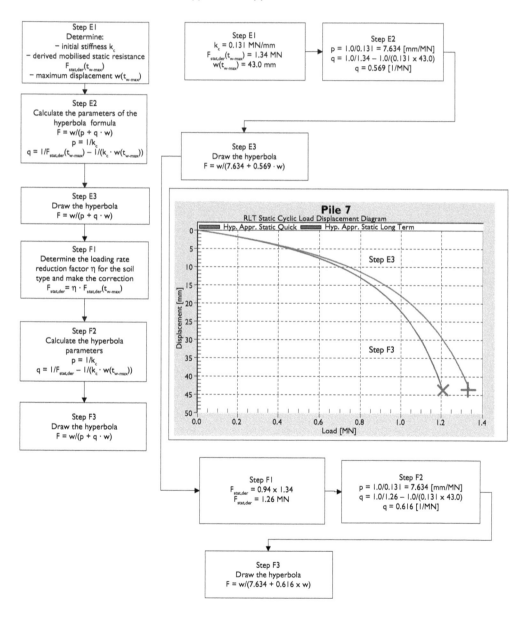

Step E1
Determine:
− initial stiffness k_c
− derived mobilised static resistance $F_{stat,der}(t_{w\text{-}max})$
− maximum displacement $w(t_{w\text{-}max})$

Step E1
$k_c = 0.131$ MN/mm
$F_{stat,der}(t_{w\text{-}max}) = 1.34$ MN
$w(t_{w\text{-}max}) = 43.0$ mm

Step E2
$p = 1.0/0.131 = 7.634$ [mm/MN]
$q = 1.0/1.34 − 1.0/(0.131 \times 43.0)$
$q = 0.569$ [1/MN]

Step E2
Calculate the parameters of the
hyperbola formula
$F = w/(p + q \cdot w)$
$p = 1/k_c$
$q = 1/F_{stat,der}(t_{w\text{-}max}) − 1/(k_c \cdot w(t_{w\text{-}max}))$

Step E3
Draw the hyperbola
$F = w/(7.634 + 0.569 \cdot w)$

Step E3
Draw the hyperbola
$F = w/(p + q \cdot w)$

Step F1
Determine the loading rate
reduction factor η for the soil
type and make the correction
$F_{stat,der} = \eta \cdot F_{stat,der}(t_{w\text{-}max})$

Step F2
Calculate the hyperbola
parameters
$p = 1/k_c$
$q = 1/F_{stat,der} − 1/(k_c \cdot w(t_{w\text{-}max}))$

Step F3
Draw the hyperbola
$F = w/(p + q \cdot w)$

Step F1
$F_{stat,der} = 0.94 \times 1.34$
$F_{stat,der} = 1.26$ MN

Step F2
$p = 1.0/0.131 = 7.634$ [mm/MN]
$q = 1.0/1.26 − 1.0/(0.131 \times 43.0)$
$q = 0.616$ [1/MN]

Step F3
Draw the hyperbola
$F = w/(7.634 + 0.616 \times w)$

Multiple cycle testing (UPM)
Overview all steps

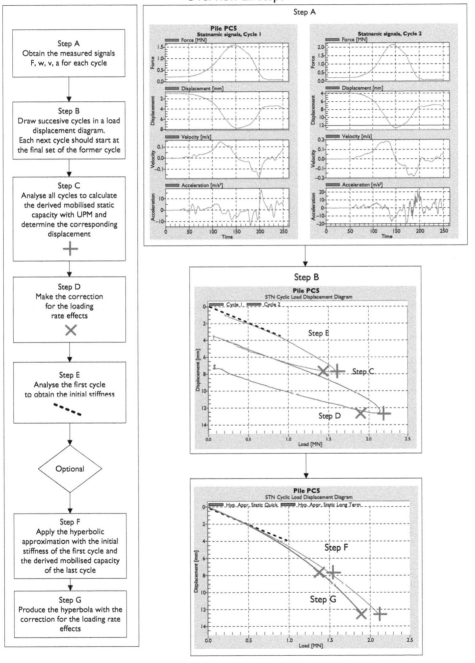

Step A
Obtain the measured signals F, w, v, a for each cycle

Step B
Draw succesive cycles in a load displacement diagram. Each next cycle should start at the final set of the former cycle

Step C
Analyse all cycles to calculate the derived mobilised static capacity with UPM and determine the corresponding displacement

Step D
Make the correction for the loading rate effects

Step E
Analyse the first cycle to obtain the initial stiffness

Optional

Step F
Apply the hyperbolic approximation with the initial stiffness of the first cycle and the derived mobilised capacity of the last cycle

Step G
Produce the hyperbola with the correction for the loading rate effects

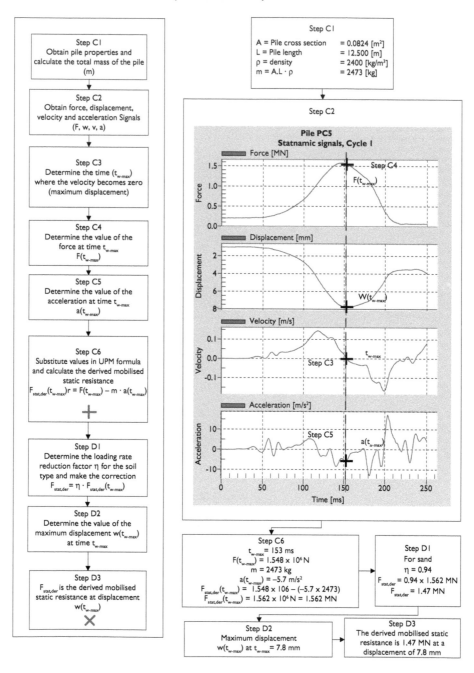

Step C1
Obtain pile properties and calculate the total mass of the pile (m)

Step C2
Obtain force, displacement, velocity and acceleration Signals (F, w, v, a)

Step C3
Determine the time (t_{w-max}) where the velocity becomes zero (maximum displacement)

Step C4
Determine the value of the force at time t_{w-max}
$F(t_{w-max})$

Step C5
Determine the value of the acceleration at time t_{w-max}
$a(t_{w-max})$

Step C6
Substitute values in UPM formula and calculate the derived mobilised static resistance
$F_{stat,der}(t_{w-max})r = F(t_{w-max}) - m \cdot a(t_{w-max})$

Step D1
Determine the loading rate reduction factor η for the soil type and make the correction
$F_{stat,der} = \eta \cdot F_{stat,der}(t_{w-max})$

Step D2
Determine the value of the maximum displacement $w(t_{w-max})$ at time t_{w-max}

Step D3
$F_{stat,der}$ is the derived mobilised static resistance at displacement $w(t_{w-max})$

Step C1

A = Pile cross section	= 0.0824 [m²]
L = Pile length	= 12.500 [m]
ρ = density	= 2400 [kg/m³]
m = A.L · ρ	= 2473 [kg]

Step C2

Pile PC5
Statnamic signals, Cycle 1

Step C6
t_{w-max} = 153 ms
$F(t_{w-max})$ = 1.548 × 10⁶ N
m = 2473 kg
$a(t_{w-max})$ = –5.7 m/s²
$F_{stat,der}(t_{w-max})$ = 1.548 × 106 – (–5.7 × 2473)
$F_{stat,der}(t_{w-max})$ = 1.562 × 10⁶ N = 1.562 MN

Step D1
For sand
η = 0.94
$F_{stat,der}$ = 0.94 × 1.562 MN
$F_{stat,der}$ = 1.47 MN

Step D2
Maximum displacement
$w(t_{w-max})$ at t_{w-max} = 7.8 mm

Step D3
The derived mobilised static resistance is 1.47 MN at a displacement of 7.8 mm

Multiple cycle testing (UPM)
Steps C and D – Cycle 2

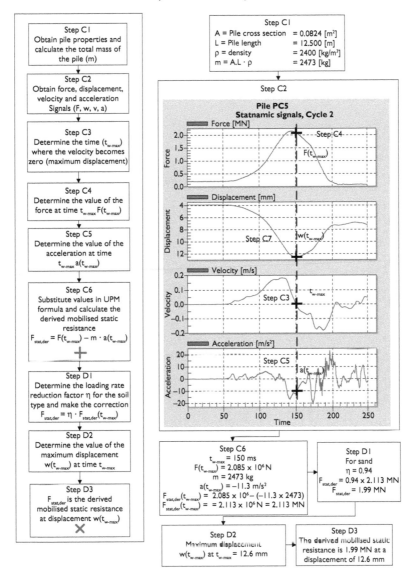

Step C1
Obtain pile properties and calculate the total mass of the pile (m)

Step C2
Obtain force, displacement, velocity and acceleration Signals (F, w, v, a)

Step C3
Determine the time (t_{w-max}) where the velocity becomes zero (maximum displacement)

Step C4
Determine the value of the force at time t_{w-max} $F(t_{w-max})$

Step C5
Determine the value of the acceleration at time t_{w-max} $a(t_{w-max})$

Step C6
Substitute values in UPM formula and calculate the derived mobilised static resistance
$F_{stat,der} = F(t_{w-max}) - m \cdot a(t_{w-max})$

Step D1
Determine the loading rate reduction factor η for the soil type and make the correction
$F_{stat,der} = η \cdot F_{stat,der}(t_{w-max})$

Step D2
Determine the value of the maximum displacement $w(t_{w-max})$ at time t_{w-max}

Step D3
$F_{stat,der}$ is the derived mobilised static resistance at displacement $w(t_{w-max})$

Step C1
A = Pile cross section = 0.0824 [m²]
L = Pile length = 12.500 [m]
ρ = density = 2400 [kg/m³]
m = A.L · ρ = 2473 [kg]

Step C2

Pile PC5
Statnamic signals, Cycle 2

Force [MN] — Step C4 — $F(t_{w-max})$

Displacement [mm] — Step C7 — $w(t_{w-max})$

Velocity [m/s] — Step C3 — t_{w-max}

Acceleration [m/s²] — Step C5 — $a(t_{w-max})$

Step C6
t_{w-max} = 150 ms
$F(t_{w-max})$ = 2.085 × 10⁶ N
m = 2473 kg
$a(t_{w-max})$ = −11.3 m/s²
$F_{stat,der}(t_{w-max})$ = 2.085 × 10⁶ − (−11.3 × 2473)
$F_{stat,der}(t_{w-max})$ = = 2.113 × 10⁶ N = 2.113 MN

Step D1
For sand
η = 0.94
$F_{stat,der}$ = 0.94 × 2.113 MN
$F_{stat,der}$ = 1.99 MN

Step D2
Maximum displacement
$w(t_{w-max})$ at t_{w-max} = 12.6 mm

Step D3
The derived mobilised static resistance is 1.99 MN at a displacement of 12.6 mm

Multiple cycle testing (UPM)
Step E
Initial stiffness

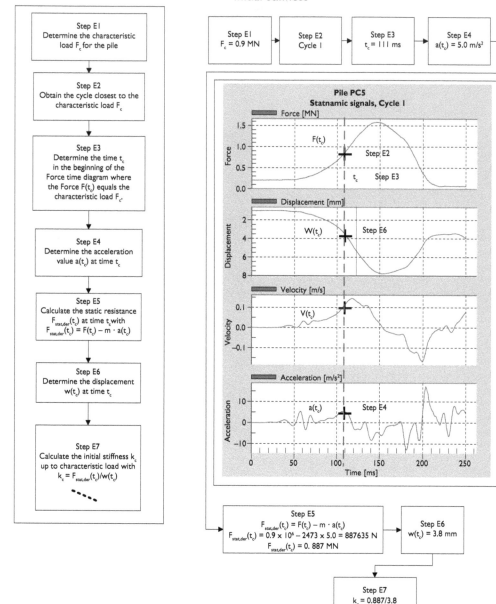

Step E1
Determine the characteristic load F_c for the pile

Step E2
Obtain the cycle closest to the characteristic load F_c

Step E3
Determine the time t_c in the beginning of the Force time diagram where the Force $F(t_c)$ equals the characteristic load F_c.

Step E4
Determine the acceleration value $a(t_c)$ at time t_c

Step E5
Calculate the static resistance $F_{stat,der}(t_c)$ at time t_c with $F_{stat,der}(t_c) = F(t_c) - m \cdot a(t_c)$

Step E6
Determine the displacement $w(t_c)$ at time t_c

Step E7
Calculate the initial stiffness k_c up to characteristic load with $k_c = F_{stat,der}(t_c)/w(t_c)$

Step E1
$F_c = 0.9$ MN

Step E2
Cycle 1

Step E3
$t_c = 111$ ms

Step E4
$a(t_c) = 5.0$ m/s^2

Pile PC5
Statnamic signals, Cycle 1

Force [MN]
$F(t_c)$ Step E2
t_c Step E3

Displacement [mm]
$W(t_c)$ Step E6

Velocity [m/s]
$V(t_c)$

Acceleration [m/s^2]
$a(t_c)$ Step E4

Time [ms]

Step E5
$F_{stat,der}(t_c) = F(t_c) - m \cdot a(t_c)$
$F_{stat,der}(t_c) = 0.9 \times 10^6 - 2473 \times 5.0 = 887635$ N
$F_{stat,der}(t_c) = 0.887$ MN

Step E6
$w(t_c) = 3.8$ mm

Step E7
$k_c = 0.887/3.8$
$k_c = 0.233$ MN/mm

Multiple cycle testing (UPM)
Steps F and G
Hyperbolic Approximation

Step F1
Detemine:
initial stiffness k_c derived mobilised
static resistance $F_{stat,der}(t_{w-max})$
maximum displacement $w(t_{w-max})$

Step F1
$k_c = 0.233$ MN/mm
$F_{stat,der}(t_{w-max}) = 2.085$ MN
$w(t_{w-max}) = 12.6$ mm

Step F2
$p = 1.0/0.233 = 4.292$ [mm/MN]
$q = 1.0/2.085 - 1.0/(0.233 \times 12.6)$
$q = 0.138$ [1/MN]

Step F2
Calculate the parameters
of the hyperbola formula
$F = w/(p + q \cdot w)$
$p = 1/kc$
$q = 1/F_{stat,der}(t_{w-max}) - 1/(k_c \cdot w(t_{w-max}))$

Step F3
Draw the hyperbola
$F = w/(4.292 + 0.138 \cdot w)$

Step F3
Draw the hyperbola
$F = w/(p + q \cdot w)$

Step G1
Determine the loading rate
reduction factor η for the soil
type and make the correction
$F_{stat,der} = η \cdot F_{stat,der}(t_{w-max})$

Step G2
Calculate the hyperbola
parameters $p = 1/k_c$
$q = 1/F_{stat,der} - 1/(k_c \cdot w(t_{w-max}))$

Step G3
Draw the hyperbola
$F = w/(p + q \cdot w)$

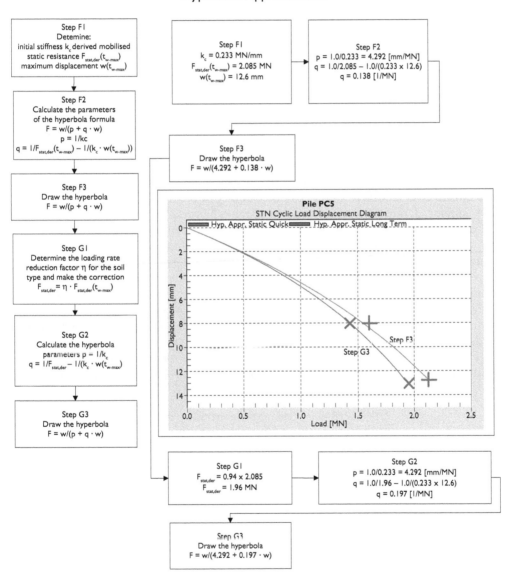

Pile PC5
STN Cyclic Load Displacement Diagram
Hyp. Appr. Static Quick — Hyp. Appr. Static Long Term
Displacement [mm]
Load [MN]
Step F3
Step G3

Step G1
$F_{stat,der} = 0.94 \times 2.085$
$F_{stat,der} = 1.96$ MN

Step G2
$p = 1.0/0.233 = 4.292$ [mm/MN]
$q = 1.0/1.96 - 1.0/(0.233 \times 12.6)$
$q = 0.197$ [1/MN]

Step G3
Draw the hyperbola
$F = w/(4.292 + 0.197 \cdot w)$

Empirical data UPM method, background

This Appendix offers background information on the empirical factor for the UPM method.

The database of McVay, Kuo et al. (2003) was used as a starting point. The results selected by McVay et al. are presented in Appendix C. Table B1 presents the resulting empirical factors η, based on this data. Three materials (clay, sand, silt) are distinguished, since the theory suggests strong differences between these materials (see Chapter 4). It is noted that the original reference by McVay et al. carried out advanced probabilistic calculations related to the probability of failure.

A similar database was constructed from other results available in literature (Hölscher and van Tol 2009b). Appendix C shows the results. A distinction between cases with geotechnical failure and without geotechnical failure has been made. The distinction is based on the maximum pile displacement during the rapid load test, for failure it must be of the order or larger than 10% of the pile diameter. Table B2 shows the statistical results for the second database. The results are quite comparable with the results from the McVay database. The standard deviation for clay is much lower, but these numbers are based on one site only with a single soil type, where 6 piles are tested.

(Hölscher and van Tol 2009b) distinguished also the method of pile installation. The main difference between installation methods is the question whether the soil

Table B1 Results statistical elaboration McVay database.

Material	Clay	Sand	Silt
Empirical factor η	0.53	0.92	0.70
Standard deviation	0.49	0.17	0.19
Number of cases	6	9	14

Table B2 Results statistical elaboration empirical database, cases with geotechnical failure.

Material	Clay	Sand
Empirical factor η	0.50	0.92
Standard deviation	0.07	0.24
Number of cases	6	12

is displaced (which leads to a higher stress in the soil) or removed (which leads to a constant or even decreasing stress in the soil). This study resulted in the following Table B3.

For piles in clay, only data for displacement piles is found in the literature. For statistical purposes, the number of sites refers to independent cases.

The IFCO building in Waddinxveen is founded on several pile types. After construction of the foundation, several piles were tested using RLT. After construction of the building, the piles were tested by an SLT. The weight of the building acted as the counter weight. Table B4 shows the results for three pile types (Middendorp 2010). The soil consists of about 8 m peat and clay, on a pleistocene sand layer. The toe of all piles is in the sand layer. The average value of the empirical coefficient η is presented in the first line, while the variation is shown in the second line. These measurements confirm the higher average of the empirical coefficient η for the CFA piles. The coefficient of variation is here lower than in Table B3, due to the fact that this data refers to one site only. However, the prefabricated driven piles also show a smaller coefficient of variation than both the in-situ constructed displacement piles and the CFA piles.

For the piles in clay, the coefficient of variation is high, as seen in Table B3. A log-normal distribution must be used. This suggests that the denomination "clay" is not sufficient. (Schmuker 2005) suggests that the empirical factor η depends on the viscosity of the clay, the higher the viscosity index, the lower the empirical factor. Schmuker suggests the formula:

$$R = \left(\frac{0.33 * 10^{-6}}{v_{max}} \right)^{I_{va}} \tag{B1}$$

Table B3 Summary.

| Pile type | Displacement | Displacement | Bored and CFA |
Soil type	Clay	Sand	Sand
Empirical factor η	0.66	0.94	1.11
Standard deviation	0.32	0.13	0.31
Coefficient of Variation	0.49	0.14	0.28
Number of cases	12	27	8
Number of sites	6	18	3

Table B4 Summary of test results foundation IFCO building.

	Displacement		
	Prefab	In-situ	CFA
Average	0.88	0.81	1.11
Standard deviation	0.09	0.21	0.23
Coefficient of variation	0.10	0.25	0.20
Number of piles	6	6	6

Where:
ν_{max} he maximum velocity reached in the test [m/s].
$I_{v\alpha}$ the viscosity index of the soil [–].

For clay, the viscosity index varies between 0.03 and 0.06, depending on the plasticity index. This equation can be used to obtain an indication of the value of the empirical factor η. Table B5 gives some values for the viscosity index, based on table 1 of (Schmuker 2005).

For piles in sand, (Huy 2008) studied the background of the empirical factor η. Based on this research, the empirical factor for the pile toe in sand turned out to have a contribution of the constitutive rate effect and was influenced by pore water pressure. For the maximum force at the pile tip, the constitutive rate effect plays a role (of about 5%), but at the unloading point this effect is negligible. The influence of pore water pressure depends on the consolidation properties of the sand around the pile toe. The effect can be expressed in the dynamic drainage factor, which can be considered as the duration of the loading relative to the duration of the consolidation, see Equation 4.5. This influence must be taken into account for piles in fine grained sands and silt.

A relatively high empirical factor with a high coefficient of variation was found for non-displacement piles, based on a very limited number of sites. The high factor suggests that the pile installation technique influences behaviour during a rapid test.

Table B5 Proposal for viscosity index.

Soil type	Sand	Silt	Clay, intermediate plasticity	Clay high plasticity	Organic clay	Peat
Viscosity index	0.01	0.02	0.03	0.04	0.06	0.07

Figure B1 Monitoring computer during rapid load testing.
Copyright: Allnamics Pile Testing Experts BV and Fugro.

Data for loading rate factor UPM

This appendix shows the data which was used to derive the empirical factor for the unloading point method.

Two tables are shown:

- The data originally analysed by McVay, Kuo et al. (2003).
- An extension of this data by the CUR committee. (Hölscher and van Tol 2009b).

Notes for the Table C1
Static test refers to a static test on the site. STN test refers to the Statnamic test on the site.

If the same pile is tested, the test order is given. The word "next" refers to the fact that the SLT and RLT are done on different, but nearby piles.

The RLT results refer to the mobilised capacity at the reference displacement. The reference displacement is taken as the lowest value of the maximum displacement reached during an RLT and an SLT. For the UPM method the results have been presented without correction for loading rate effects.

Notes for the Table C2
Two tables are presented: one with geotechnical failure (where the maximum displacement was larger than 5% of (equivalent) pile diameter; and one without geotechnical failure. The original references mentioned in the first column can be found in (Hölscher and van Tol 2009b).

Table C1 Overview of field test results from McVay database (McVay et al., 2003).

No.	Location	Pile type	Soil type	Test order	Static test kN	STN test kN
Sand an clay, acc. Huy 2004						
1	Noto, JPN	Steel pipe	soft rock	STL – STN	4380	5087
2	BC pier 5, USA	Driven PC	sand	STN – STL	3500	3957
3	BC pier 10, USA	Driven PC	sand	STL – STN	3380	5000
4	BC pier 15, USA	Driven PC	sand	STL – STN	3820	3322
5	Shonan T5, JPN	Driven bored	sand	STL – STN	446	489
6	Shonan T6, JPN	Driven pipe	sand	STN – STL	1100	1042
7	Contraband T114, USA	Driven PC	clay	STN – STL	1830	3070
8	Nia TP 5&6B, USA	Pipe	clay	STL – STN	2190	2600
9	Amherst 2, USA	Driven steel	clay	next	1214	1244
10	Amherst 4, USA	Driven steel	clay	next	965	1617
11	S9004 T1, CAN	AC	sand	next	1310	1350
12	S9102 T2, CAN	Pipe	clay	next	1040	2550
13	S9209 T1, USA	Driven steel	sand	STL – STN	7130	6370
14	S9306 T2, USA	Pipe	clay	next	1360	892
15	YKN - 5, JPN	Driven PC	sand	STL – STN	2770	2700
16	Hasaki - 6, JPN	Pipe	sand	STL – STN	1890	1490
Silty sites						
	ashaft10	DS	silt	STL-STN	1420	2530
	ashaft8	DS	silt	STL-STN	1700	1680
	ashaft7	DS	silt	STL-STN	2230	2430
	ashaft5	DS	silt	STL-STN	2800	2230
	ashaft3	DS	silt	STL-STN	1013	1200
	ashaft2	DS	silt	STL-STN	2230	2030
	ashaft1	DS	silt	STL-STN	2400	2050
	NIA TP 12a	pipe	silt	next	1230	1285
	NIA TP 12b	pipe	sil	next	1300	950
	NIA TP 13a	pipe	silt	next	1210	1225
	NIA TP 13b	pipe	silt	next	1300	1136
	NIA TP 910a	pipe	silt	next	1810	1900
	NIA TP 910b	pipe	silt	next	2380	1890
	S9010T1	DP	silt	STL-STN	2470	2360

DP driven pile DS drilled shaft.

Table C2 Overview of field test result for cases *with* geotechnical failure during the test, piles in clay.

Source	Location	Pile type	Soil type	Test order	SLT kN	SLT kN	Factor [-]
McVay et al., 2003	Contraband T114, USA	Driven PC	clay	RLT - STL	1830	3070	0.60
McVay et al., 2003	Nia TP 5 & 6B, USA	Pipe	clay	SLT - RLT	2190	2600	0.84
McVay et al., 2003	Amherst 2, USA	Driven steel	clay	next	1214	1244	0.98
McVay et al., 2003	Amherst 4, USA	Driven steel	clay	next	965	1617	0.60
McVay et al., 2003	S9102 T2, CAN	Pipe	clay	next	1040	2550	0.41
McVay et al., 2003	S9306 T2, USA	Pipe	clay	next	1360	892	1.52
Holeyman et al., 2001	Sint-Katelijne-Waver B	Prefab	Clay	nearby pile	1364	3110	0.44
Holeyman et al., 2001	Sint-Katelijne-Waver B	Fundex	Clay	nearby pile	1216	3033	0.40
Holeyman et al., 2001	Sint-Katelijne-Waver B	de Waal	Clay	nearby pile	1258	2580	0.49
Holeyman et al., 2001	Sint-Katelijne-Waver B	Olivier	Clay	nearby pile	1722	3100	0.56
Holeyman et al., 2001	Sint-Katelijne-Waver B	Omega	Clay	nearby pile	1263	2454	0.51
Holeyman et al., 2001	Sint-Katelijne-Waver B	Atlas	Clay	Nearby pile	1637	2766	0.59

Table C3 Overview of field test result for cases *with* geotechnical failure during the test, displacement piles in sand.

Source	Location	Pile type	Soil type	Test order	SLT kN	SLT kN	Factor [-]
McVay et al., 2003	BC pier 5, USA	Driven PC	sand	RLT – SLT	3500	3957	0.88
McVay et al., 2003	BC pier 10, USA	Driven PC	sand	SLT – RLT	3380	5000	0.68
McVay et al., 2003	BC pier 15, USA	Driven PC	sand	SLT – RLT	3820	3322	1.15
McVay et al., 2003	Shonan T5, JPN	Driven bored	sand	SLT – RLT	446	489	0.91
McVay et al., 2003	Shonan T6, JPN	Driven pipe	sand	RLT – SLT	1100	1042	1.06
McVay et al., 2003	S9004 T1, CAN	AC	sand	next	1310	1350	0.97
McVay et al., 2003	S9209 T1, USA	Driven steel	sand	SLT – RLT	7130	6370	1.12
McVay et al., 2003	YKN – 5, JPN	Driven PC	sand	SLT – RLT	2770	2700	1.03
McVay et al., 2003	Hasaki – 6, JPN	Pipe	sand	SLT – RLT	1890	1490	1.27
Heijnen, Joustra 1996	Delft, NL, pile 5	concrete driven	clay with sand layers, tip in sand	SLT – RLT	1200	1800	0.67
Opstal et al., 1996	Maasvlakte NL, pile 1	steel HP heavy point, open end	sand with clay layers	unknown	5350	5800	0.92
Opstal et al., 1996	Maasvlakte NL, pile 2	steel HP heavy point, closed end	sand with clay layers	unknown	6500	7100	0.92
Opstal et al., 1996	Maasvlakte NL, pile 3	steel HP heavy point, closed end	sand with clay layers	unknown	6400	7500	0.85

(Continued)

Table C3 (Continued).

Source	Location	Pile type	Soil type	Test order	SLT kN	SLT kN	Factor [-]
Opstal et al., 1996	Maasvlakte NL, pile 6	concrete with casing	sand with clay layers	unknown	4400	5000	0.88
Opstal et al., 1996	Maasvlakte NL, pile 8	concrete with casing	sand with clay layers	unknown	4630	4800	0.96
Opstal et al., 1996	Maasvlakte NL, pile 10	concrete with casing	sand with clay layers	unknown	4290	5100	0.84
Holeyman et al., 2003	Limelette B	de Waal	top clay, deep sand	nearby pile	2200	2850	0.77
Holeyman et al., 2003	Limelette B	Fundex	top clay, deep sand	nearby pile	2850	3000	0.95
Holeyman et al., 2003	Limelette B	Omega	top clay, deep sand	nearby pile	2700	2550	1.06
Middendorp et al., 1992	Utrecht NL	precast tubular drilled	sand	nearby pile	1060	1100	0.96
Middendorp et al., 1992	Utrecht NL	precast tubular drilled	sand	nearby pile	1060	1200	0.88
Hölscher et al., 2009	Waddinxveen NL, Pile 1	concrete driven	sand with clay layers	RLT – SLT	1150	1230	0.93

Table C4 Overview of field test result for cases *with* geotechnical failure during the test, bored and CFA piles in sand.

Source	Location	Pile type	Soil type	Test order	SLT kN	SLT kN	Factor
Briaud 2000	sand, pile 2	bored pile with defects	Sand	rapid, (dynamic,) static	1602	2460	0.65
Presten 2001	mixed, pile site 1 middle	drilled pile	sand/clay layers	rapid, static	5200	3300	1.58
Holeyman et al., 2003	Limelette B	de Waal	top clay, deep sand	nearby pile	2200	2850	0.77
Holeyman et al., 2003	Limelette B	Fundex	top clay, deep sand	nearby pile	2850	3000	0.95
Holeyman et al., 2003	Limelette B	Omega	top clay, deep sand	nearby pile	2700	2550	1.06
Middendorp 2010	Waddinxveen pile 48	CFA	clay, toe in sand	RLT – SLT	920	742	1.24
Middendorp 2010	Waddinxveen pile 49	CFA	clay, toe in sand	RLT – SLT	840	1221	0.69
Middendorp 2010	Waddinxveen pile 50	CFA	clay, toe in sand	RLT – SLT	820	649	1.26
Middendorp 2010	Waddinxveen pile 51	CFA	clay, toe in sand	RLT – SLT	540	503	1.07
Middendorp 2010	Waddinxveen pile 52	CFA	clay, toe in sand	RLT – SLT	840	754	1.11
Middendorp 2010	Waddinxveen pile 53	CFA	clay, toe in sand	RLT – SLT	840	653	1.29

Table C5 Overview of field test result for cases *without* geotechnical failure during the test.

Source	Location	Pile type	Soil type	Test order	SLT kN	RLT kN
Middendorp 1992	Embden BRD	Franki	Clay, sand, loam	rapid, static	3750	4020
Middendorp 1992	Embden BRD	Franki	clay	rapid, static	3150	3390
Horvath et al., 2004	New York USA	tapered steel with concrete	sand with clay layer	unknown	850	800
Briaud 2000	NGES-TAMU, pile 4	bored pile	sand	rapid, (dynamic,) static	4004	4490
Briaud 2000	NGES-TAMU, pile 7	bored pile	clay	rapid, (dynamic,) static	3025	3150
Brown 2004	Grimsby, in situ pile, UK	auger bored pile	clay	rapid, static	2000	2346
Hajduk 1998	Newburry USA, TP 2	driven steel pipe (open?)	clay	static, rapid	800	1200
Hajduk 1998	Newburry USA, TP 3	driven concrete	clay	static, rapid	1025	1450
Opstal et al., 1996	Maasvlakte NL, pile 1	steel HP heavy point, open end	sand with clay layers	unknown	3250	3800
Opstal et al., 1996	Maasvlakte NL, pile 2	steel HP heavy point, closed end	sand with clay layers	unknown	3200	3700
Opstal et al., 1996	Maasvlakte NL, pile 3	steel HP heavy point, closed end	sand with clay layers	unknown	2700	3050
Opstal et al., 1996	Maasvlakte NL, pile 6	concrete with casing	sand with clay layers	unknown	2900	3300
Opstal et al., 1996	Maasvlakte NL, pile 8	concrete with casing	sand with clay layers	unknown	3200	2900
Opstal et al., 1996	Maasvlakte NL, pile 10	concrete with casing	sand with clay layers	unknown	3050	2800
Holeyman et al., 2003	Limelette B	Prefab	top clay, deep sand	nearby pile	2800	3400
Holeyman et al., 2003	Limelette B	Atlas	top clay, deep sand	nearby pile	2850	3400
Holeyman et al., 2003	Limelette B	Olivier	top clay, deep sand	nearby pile	2000	1750
Middendorp et al., 2008	P2, Minden BRD	Jacbo SOB	clay	statnamic, static	3400	4100
Middendorp et al., 2008	P12, Minden BRD	Jacbo SOB	clay	statnamic, static	3600	4500
M.C.Romano et al., 2010	Maasvlakte NL SLT-PC7 STN-PC5	Vibro pile	sand	nearby pile	1900	2100
M.C.Romano et al., 2010	Maasvlakte NL SLT-P10 STN-PC11	Vibro pile	sand	nearby pile	2850	2900
M.C.Romano et al., 2010	Maasvlakte NL SLT-PC8 STN-PC6	Vibro pile	sand	nearby pile	2700	3250
M.C.Romano et al., 2010	Maasvlakte NL SLT-PC9 STN-PC12	Vibro pile	sand	nearby pile	3900	4000
Hölscher et al., 2009	Waddinxveen NL, Pile 2	Concrete driven	sand with clay layers	SLT-RLT	1050	1210

Step by step description of the Sheffield Method (SHM)

For the Sheffield method, the following step-by step approach is described. This step-by-step method is related to Section 5.3 of the guideline.

Step A   Obtain measured and derived signals; Force, displacement, velocity and acceleration.
Step B   Calculate the derived static load displacement diagram.
Step B1   Obtain the pile properties and calculate the pile mass.
Step B2   Determination of material parameter.

The parameters α and β, which describe the rate dependency of the fine-grained soil, must be determined. Three methods are available. The choice depends on the availability of soil data. Method 1a is preferred, method 1b is only useful if a proper soil classification is available.

Method 1a: from a laboratory test
A penetration test or a fast-triaxial test at several displacement rates must be executed. Tests should be performed at a minimum of five displacement rates to reflect the velocity content of a full scale rapid load test in the same soil type. The measured resistance at elevated displacement rates is normalised by the resistance at the lowest displacement rate (v_{static}) and is then presented as a function of dimensionless loading velocity (which is obtained from division of the actual loading velocity for each test by a reference velocity $v_{ref} = 1$ m/s). The lowest displacement rate test should be selected to reflect in situ full scale static pile testing rates and should be appropriately scaled if drained conditions are encountered. Note that a set of parameters α and β without a known value of v_{static} or v_{ref} is useless.

Method 1b: from literature
Here: $v_{static} = 1*10^{-5}$ m/s = 0.01 mm/s and $v_{ref} = 1$ m/s = 1000 mm/s is chosen. Note: the value α depends on both the parameters v_{static} and v_{ref} used for the interpretation of the test results, v_o cannot be freely chosen.

Method 1c: from rule of thumb
   $\beta = 0.2$ and $\alpha = 0.031*PI + 0.46$

Where PI is the Plasticity Index of the clay [%]

Table D1 Suggested values.

Originator	Soil	Index properties (LL, PL, PI-%)	α [-]	β [-]	Test conditions
Randolph & Deeks	Sand	–	0.1	0.2	Summary of previous
	Clay	–	1.0	0.2	work
Balderas-Meca	Grimsby glacial till	20–36, 12–18, 7–20	0.9	0.2	Statnamic tests
Brown	Model clay	37, 17, 20	1.26	0.34	Model Statnamic tests
Poskitt & Leonard	Cowden till	40, 20, 20	1.0	0.27	Penetrometer tests
Litkouhi & Poskitt	London Clay	70, 27, 43	1.77	0.18	Penetrometer tests
	Forties clay	30, 20, 18	0.99	0.23	
	Magnus clay	31, 17, 14	0.86	0.46	

Step B3: Calculate the derived static resistance

The uncorrected soil force is the total soil resistance, including the rate effects (see note on notation and calculation of velocity at the end of the appendix):

Step B3-1: Correct the measured force for inertia

$$F_{soil;i} = F_{head;i} - m\,a_i \qquad\qquad (D1)$$

Where:

$F_{soil,i}$ (calculated) uncorrected soil force on the pile at time $t_i = i*\Delta t$ [N].
$F_{head,i}$ (measured) force on the pile head at time $t_i = i*\Delta t$ [N].
m mass of the pile. This value can be calculated from the dimensions of the pile and the volumetric mass of the material [kg].
a_i (measured) acceleration of the pile at time $t_i = i*\Delta t$ [m/s²].

the subscript i is a counter over the time steps.

The index notation (where the index refers to time step) is adopted in this appendix.

Step B3-2: Calculate the correction factor for each time step

The correction factor for the rate effects is calculated for each time-step. The method is valid for the branch of the rapid load displacement until the unloading point (so downwards movement only).

$$H_i = 1 + \alpha \left(\frac{F_{head;i}}{F_{max}} \right) \left[\left(\frac{\tilde{v}_i}{v_{ref}} \right)^\beta - \left(\frac{v_{static}}{v_{ref}} \right)^\beta \right] \qquad\qquad (D3)$$

Where:

H_i dimensionless correction factor at time $t_i = (i-1)*\Delta t$ [–].
F_{max} maximum value of the force on the pile head, *i.e.* max $(F_{head,i})$ [N].

$F_{head,I}$ force on the pile head on time t_i [N].

α and β constants determined in step 1 [–]

$v_{static} = 1*10^{-5}$ m/s.

$v_{ref} = 1$ m/s.

$\tilde{v}_i$ modified pile velocity, defined by

$$\tilde{v}_i = v_{static} \quad \text{if} \quad |v_i| < v_{static}$$
$$\tilde{v}_i = |v_i| \quad \text{if} \quad t_i < t_{max}$$
$$\tilde{v}_i = v_{max} \quad \text{if} \quad t_i \geq t_{max}$$

Where:

v_{max} maximum downward velocity of the pile before the upward motion starts [m/s].

t_{max} time at which v_{max} is reached [s].

It is noted that H_i must be calculated for each time step since H_i is a function of velocity v_i and force on the pile head $F_{head,i}$, which are changing continuously during the test.

Step B3-3: Calculate the corrected soil force

The derived soil-force is calculated from

$$F_{stat,der,i} = \frac{F_{soil;i}}{H_i} \tag{D4}$$

Where:

$F_{stat,der,i}$ derived static soil force on the pile at time t_i [N].

$F_{soil,i}$ (calculated) uncorrected soil force on the pile at time t_i calculated in step B3-1 [N].

H_i correction factor at time t_i calculated in step B3-2 [–].

Step B3-4: Draw the derived load displacement curve

The curve w_i against $F_{derived,i}$ is the final result of the calculation

Where:

$F_{derived,i}$ derived static soil force on the pile at time t_i calculated in step B3-4 [N].

w_i (measured) displacement of the pile at time $t_i = (i-1)*\Delta t$ [m].

Step C: Determine maximum mobilised capacity and the displacement at which the maximum mobilised capacity was reached

Step D: Derivation of stiffness

The stiffness of the pile can be derived from the initial part of the curve. For a cyclic test, this must be done for the first cycle.

From the last cycle in the test, the mobilised static resistance can be read. This can be considered as (a lower limit of) the ultimate capacity.

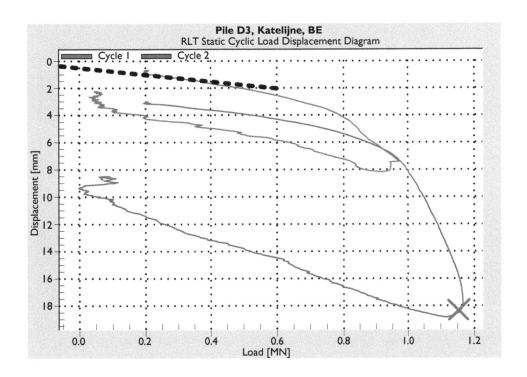

Note on notation and calculation of the velocity

The velocity of the pile can be calculated by integration of the measured acceleration by numerical integration, e.g.:

$$v_{i+1} = v_i + \tfrac{1}{2}\Delta t * (a_{i+1} + a_i)$$

Where:

v_i (calculated) velocity of the pile at time $t_i = i * \Delta t$ [m/s].
a_i (measured) acceleration of the pile at time $t_i = * \Delta t$ [m/s^2].
Δt time-step in the measurement. This time step is generally constant [s].

Note: The velocity can also be found by differentiation of the measured displacements, but generally, the method of integration is believed to be more accurate.

Sheffield Method
(SHM)
Step by step

Determine
Testing Type

Single cycle
testing

Multiple
cycle testing

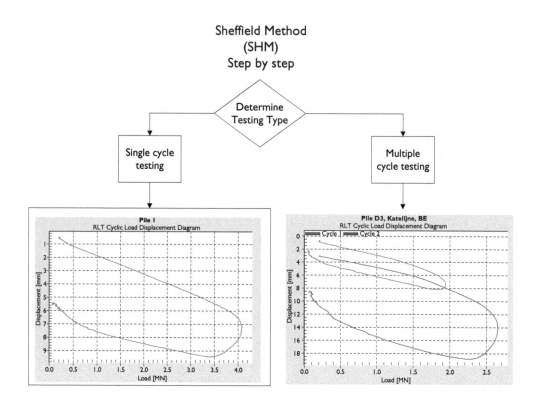

Single cycle testing (SHM)
Overview all steps

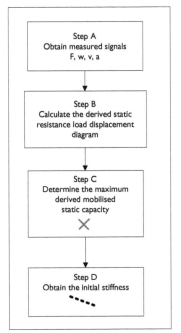

Step A
Obtain measured signals
F, w, v, a

Step B
Calculate the derived static
resistance load displacement
diagram

Step C
Determine the maximum
derived mobilised
static capacity
✕

Step D
Obtain the initial stiffness

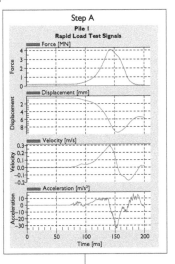

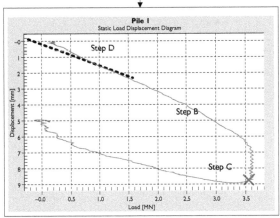

Single cycle testing (SHM),
Steps B and C.
Mobilised static capacity

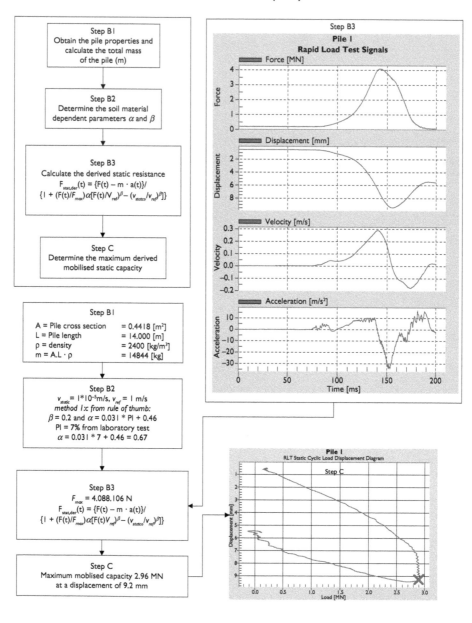

Step B1
Obtain the pile properties and calculate the total mass of the pile (m)

Step B2
Determine the soil material dependent parameters α and β

Step B3
Calculate the derived static resistance
$F_{stat,der}(t) = \{F(t) - m \cdot a(t)\}/$
$\{1 + (F(t)/F_{max})\alpha[F(t)/V_{ref})^{\beta} - (v_{statco}/v_{ref})^{\beta}]\}$

Step C
Determine the maximum derived mobilised static capacity

Step B1

A = Pile cross section	= 0.4418 [m²]
L = Pile length	= 14.000 [m]
ρ = density	= 2400 [kg/m³]
m = A.L · ρ	= 14844 [kg]

Step B2
v_{static} = 1*10⁻⁵m/s, v_{ref} = 1 m/s
method 1:x from rule of thumb:
β = 0.2 and α = 0.031 * PI + 0.46
PI = 7% from laboratory test
α = 0.031 * 7 + 0.46 = 0.67

Step B3
F_{max} = 4.088.106 N
$F_{stat,der}(t) = \{F(t) - m \cdot a(t)\}/$
$\{1 + (F(t)/F_{max})\alpha[F(t)V_{ref})^{\beta} - (v_{statco}/v_{ref})^{\beta}]\}$

Step C
Maximum moblised capacity 2.96 MN at a displacement of 9.2 mm

Single cycle testing (SHM)
Steps D
Initial stiffness

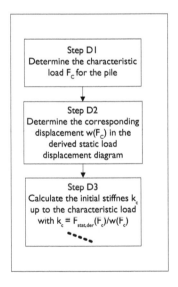

Step D1
Determine the characteristic load F_c for the pile

Step D2
Determine the corresponding displacement $w(F_c)$ in the derived static load displacement diagram

Step D3
Calculate the initial stiffnes k_c up to the characteristic load with $k_c = F_{stat,der}(F_c)/w(F_c)$

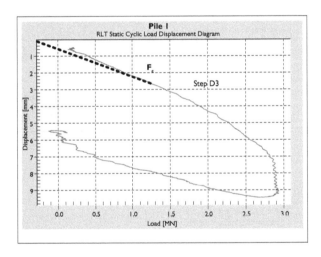

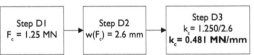

Step D1
$F_c = 1.25$ MN

Step D2
$w(F_c) = 2.6$ mm

Step D3
$k_c = 1.250/2.6$
$k_c = 0.481$ **MN/mm**

Multiple cycle testing (SHM)
Overview all steps

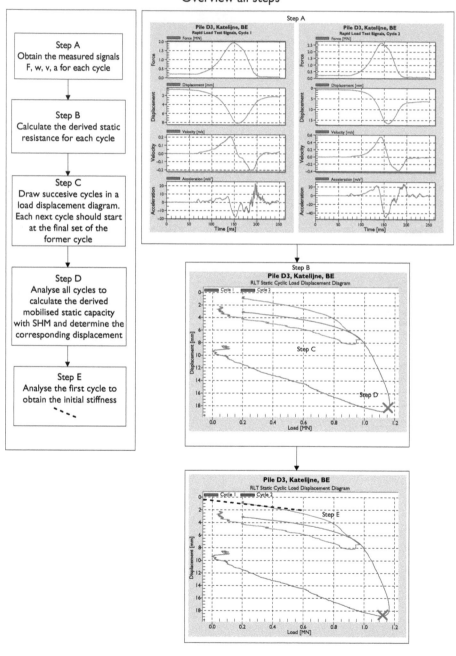

Step A
Obtain the measured signals F, w, v, a for each cycle

Step B
Calculate the derived static resistance for each cycle

Step C
Draw succesive cycles in a load displacement diagram. Each next cycle should start at the final set of the former cycle

Step D
Analyse all cycles to calculate the derived mobilised static capacity with SHM and determine the corresponding displacement

Step E
Analyse the first cycle to obtain the initial stiffness

Multiple cycle testing (SHM)
Steps B – Cycle 1

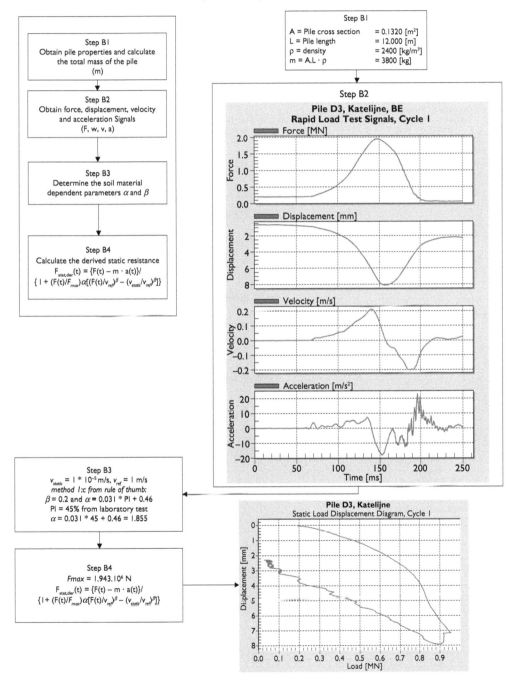

Step B1
Obtain pile properties and calculate the total mass of the pile
(m)

Step B2
Obtain force, displacement, velocity and acceleration Signals
(F, w, v, a)

Step B3
Determine the soil material dependent parameters α and β

Step B4
Calculate the derived static resistance
$F_{stat,der}(t) = \{F(t) - m \cdot a(t)\}/$
$\{1 + (F(t)/F_{max})\alpha[(F(t)/v_{ref})^{\beta} - (v_{static}/v_{ref})^{\beta}]\}$

Step B1

A = Pile cross section	= 0.1320 [m²]
L = Pile length	= 12.000 [m]
ρ = density	= 2400 [kg/m³]
m = A.L · ρ	= 3800 [kg]

Step B2

Pile D3, Katelijne, BE
Rapid Load Test Signals, Cycle 1

Force [MN]

Displacement [mm]

Velocity [m/s]

Acceleration [m/s²]

Step B3
$v_{static} = 1 * 10^{-5}$ m/s, $v_{ref} = 1$ m/s
method 1:c from rule of thumb:
$\beta = 0.2$ and $\alpha = 0.031 *$ PI $+ 0.46$
PI = 45% from laboratory test
$\alpha = 0.031 * 45 + 0.46 = 1.855$

Step B4
Fmax $= 1.943.10^6$ N
$F_{stat,der}(t) = \{F(t) - m \cdot a(t)\}/$
$\{1 + (F(t)/F_{max})\alpha[(F(t)/v_{ref})^{\beta} - (v_{static}/v_{ref})^{\beta}]\}$

Pile D3, Katelijne
Static Load Displacement Diagram, Cycle 1

Multiple cycle testing (SHM)
Steps B – Cycle 2

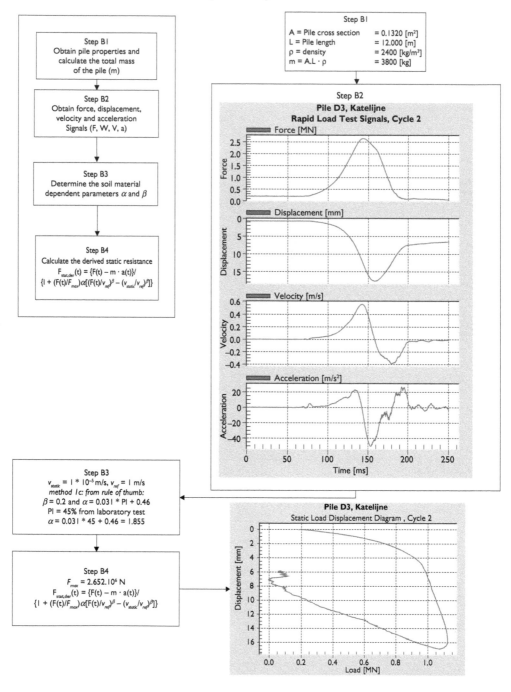

Step B1
Obtain pile properties and calculate the total mass of the pile (m)

Step B2
Obtain force, displacement, velocity and acceleration Signals (F, W, V, a)

Step B3
Determine the soil material dependent parameters α and β

Step B4
Calculate the derived static resistance
$F_{stat,der}(t) = \{F(t) - m \cdot a(t)\}/$
$\{1 + (F(t)/F_{max})\alpha[(F(t)/v_{ref})^\beta - (v_{static}/v_{ref})^\beta]\}$

Step B1

A = Pile cross section	= 0.1320 [m²]
L = Pile length	= 12.000 [m]
ρ = density	= 2400 [kg/m³]
m = A.L · ρ	= 3800 [kg]

Step B2
Pile D3, Katelijne
Rapid Load Test Signals, Cycle 2

Step B3
$v_{static} = 1 * 10^{-5}$ m/s, $v_{ref} = 1$ m/s
method 1c: from rule of thumb:
$\beta = 0.2$ and $\alpha = 0.031 * PI + 0.46$
PI = 45% from laboratory test
$\alpha = 0.031 * 45 + 0.46 = 1.855$

Step B4
$F_{max} = 2.652.10^6$ N
$F_{stat,der}(t) = \{F(t) - m \cdot a(t)\}/$
$\{1 + (F(t)/F_{max})\alpha[(F(t)/v_{ref})^\beta - (v_{static}/v_{ref})^\beta]\}$

Pile D3, Katelijne
Static Load Displacement Diagram , Cycle 2

Multiple cycle testing (SHM)
Step C and D

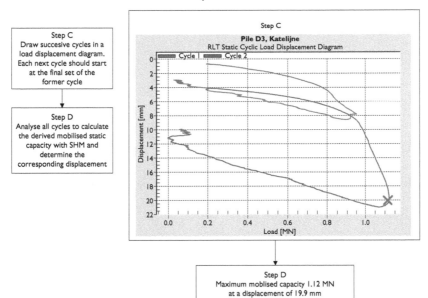

Step C
Draw succesive cycles in a load displacement diagram. Each next cycle should start at the final set of the former cycle

Step D
Analyse all cycles to calculate the derived mobilised static capacity with SHM and determine the corresponding displacement

Step D
Maximum moblised capacity 1.12 MN at a displacement of 19.9 mm

Multiple cycle testing (SHM)
Steps
EInitial stiffness

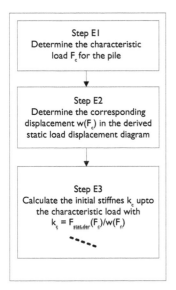

Step E1
Determine the characteristic load F_c for the pile

Step E2
Determine the corresponding displacement $w(F_c)$ in the derived static load displacement diagram

Step E3
Calculate the initial stiffnes k_c upto the characteristic load with
$k_c = F_{stat,der}(F_c)/w(F_c)$

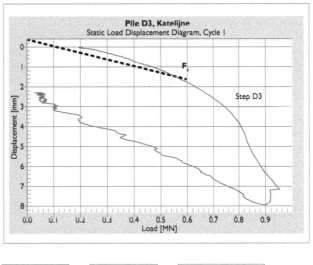

Step D1
$F_c = 0.60$ MN

Step D2
$w(F_c) = 1.8$ mm

Step D3
$k_c = 0.6/1.8$
$k_c = 0.333$ MN/mm

Appendix E

Description of rate effects in clay

E.1 AVAILABLE METHODS

In this appendix, three descriptions of the rate effect in clay are shortly compared.

1   The power law used by e.g. (Schmuker 2004)

$$\frac{\tau}{\tau_s} = \left(\frac{v}{v_s} \right)^{\beta}$$ (E.1)

Where:
τ strength [Pa]
τ_s strength at velocity v_s [Pa]
v velocity [m/s]
v_s 'static' velocity [m/s]
β material parameter [-]
2   The equation used by e.g. (Randolph 1992)

$$\frac{\tau}{\tau_s} = 1 + \alpha \left(\frac{v}{v_{ref}} \right)^{\beta}$$ (E.2)

Where:
τ strength [Pa]
τ_s strength at velocity v_o [Pa]
v velocity [m/s]
v_{ref} reference velocity [m/s]
α material parameter [-]
β material parameter [-]
3   The equation from the Sheffield group (Brown 2004)

$$\frac{\tau}{\tau_s} = 1 + \alpha \left[\left(\frac{v}{v_{ref}} \right)^{\beta} - \left(\frac{v_s}{v_{ref}} \right)^{\beta} \right]$$ (E.3)

Where:

τ strength [Pa].
τ_s strength at velocity v_s [Pa].
v velocity [m/s].
v_{ref} reference velocity [m/s].
v_s 'static' velocity [m/s].
α material parameter [–].
β material parameter [–].

E.2   COMPARISONS

To show the essential differences between the three methods, the data shown in the following table E1 (velocities in m/s) are used. Three curves are drawn in a log-log graph.

It shows that the power law gives a behaviour at low velocity, which has no static limit. This seems unrealistic. Based on physical background and experience, an increase of the rate effect with increasing loading rate seems reasonable. The difference between the methods called Randolph and Sheffield are much smaller. The Sheffield

Table E1 Comparison of methods.

	Schmuker	Randolph	Brown
V_{ref}	1	1	1
V_{stat}			0,01
α		1	1
β	0.2	0.2	0.2

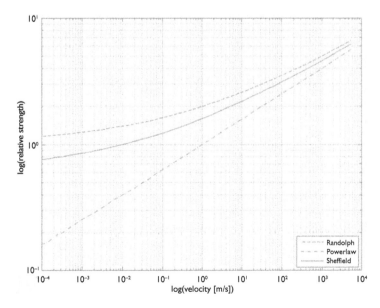

Figure E1 Comparison common descriptions of rate effect over large range of loading velocity.

method has the advantage that the term 'static' is properly defined. This is not the case in the equation proposed by Randolph (formally, this graph cannot be drawn for the Randolph equation, since the relative strength depends on the strength at zero speed).

In practical usage the static reference velocity is chosen much smaller. At a smaller static reference velocity the differences between the Randolph method and the Sheffield method are smaller. For three choices of the static velocity the curves are drawn, together with the curve suggested by Randolph. These parameters cannot be exchanged without correction.

Finally, it is remarked that over a limited range of velocity the power law can be used as a linearization of the rate effect. It must be clearly noted that the parameters used for that linearization are valid only for a smaller velocity range.

E.3   UNIQUENESS OF PARAMETERS

Lets assume for instruction that the Sheffield equation is the most reliable description of the phenomenon. This shouldn't be true absolutely, but it seems reasonable that the velocity which is used in physical testing is not the physical lower limit (as assumed by Randolph) and it seems reasonable that the rate effect is not constant over a very large range of velocities. Physical tests over a wide range of velocities must give final evidence.

This explains why the usage of the power law leads to an unexpected low rate effect. The data presented by Schmuker (2005) are derived from very slow tests (10^{-5} to 10^{-3}). Extrapolation to the velocity range of interest at RLT gives indeed a much lower velocity effect than expected from the Sheffield equation.

The graph also explains why we should be very accurate and cautious with the usage of the reference velocity and static velocity. In fact the Randolph equation is

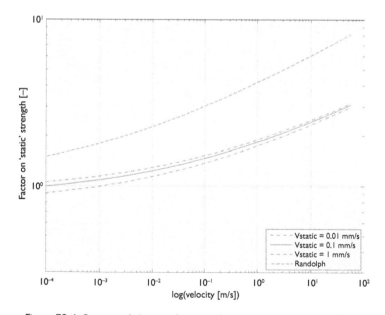

Figure E2 Influence of choice of static velocity on derived rate effect.

not solvable since we cannot do a test with infinite low velocity. That means that the Randolph equation requires always extrapolation to infinite low velocity, which may result in unambiguity of the interpretation.

Finally, the reference velocity v_o cannot be chosen freely (by both the approach of Randolph and Sheffield). This can be shown by a "reductio ad adsurdum". Thus:

$$\left(\frac{v}{v_1}\right)^\beta - \left(\frac{v_{s1}}{v_1}\right)^\beta = \left(\frac{v}{v_2}\right)^\beta - \left(\frac{v_{s2}}{v_2}\right)^\beta \tag{E4}$$

Assume $v_2 = R\,v_1$ this equation reduces to

$$\left(\frac{v}{v_1}\right)^\beta - \left(\frac{v_{s1}}{v_1}\right)^\beta = \frac{1}{R^\beta}\left[\left(\frac{v}{v_1}\right)^\beta - \left(\frac{v_{s2}}{v_1}\right)^\beta\right] \tag{E5}$$

This must be true for all velocities v, which requires $R = 1$ and $v_{s1} = v_{s2}$.

This shows that the assumption that α and β are independent of the reference value leads to the fact that this is only possible for one reference value.

The following conclusions can be drawn:

– Care should be taken for the proper use of the reference value and static value of the velocity.
– The values of α and β do not make sense without the reference and static values.
– The power law describes the velocity dependency over a limited range of velocities.

Figure E3 Transport of a 4 MN rapid load test device on the building site.
Copyright: Allnamics Pile Testing Experts BV and Profound BV.

Appendix F

Example of safety approach

The test with concrete piles in sand layers in Waddinxveen (Hölscher, Brassinga et al., 2009) is interpreted as an example in Appendix A. This example will be elaborated in this Appendix.

The number of piles tested was 2. The derived static ultimate bearing capacity (using UPM) was determined to be 1.14 MN and 1.13 MN.

1 Determine characteristic value as the minimum from:

$$R_{c;k} = Min \left\{ \frac{R_{c;RLT;avg}}{\xi_{1;RLT}}, \frac{R_{c;RLT;min}}{\xi_{2;RLT}} \right\} \tag{F1}$$

Where:

$$\xi_{1,RLT} = \frac{1}{2}(\xi_{1;SLT} + 0.85 \cdot \xi_{5;DLT}) \tag{F2}$$

$$\xi_{2,RLT} = \frac{1}{2}(\xi_{2;SLT} + 0.85 \cdot \xi_{6;DLT}) \tag{F3}$$

The average value ($R_{c;RLT;avg}$) is 1.14 MN and the minimum value ($R_{c;RLT;min}$) is 1.13 MN. The correlation factors of SLT for a stiff building are 1.32 for $\xi_{1;SLT}$ and $\xi_{2;SLT}$. The correlation factors of DLT are 1.6 for $\xi_{5;DLT}$ and 1.5 for $\xi_{6;DLT}$.

$$R_{c;k} = Min \left\{ \frac{1.14}{1.34}, \frac{1.13}{1.30} \right\} = Min\{0.85; 0.87\} = 0.85 \text{ MN}$$

2 Calculate the design value follows from:

$$R_{c;d} = \frac{R_{c;k}}{\gamma_R} = \frac{0.85}{1.15} = 0.74 \text{ MN}$$

At Waddinxveen also an SLT was undertaken. The bearing capacity at the same pile settlement as the RLT was about 1.15 MN for Pile 1 and 1.12 MN for Pile 2. The characteristic value will be $1.12/1.32 = 0.85$. The design value will be $0.85/1.15 = 0.74$. This value is comparable with the result of the RLT.

Appendix G

Database

G.1 INTRODUCTION

A database for calibration purposes is essential to judge the reliability of interpretation methods. In this chapter the requirement for such data is described. The presented empirical cases contain measurement of data for both rapid load tests and static load test on the same or a neighbouring pile.

The database, presented in Appendix C, contains a relatively large number of cases where UPM analysis has previously been carried out. This facilitates a general statistical approach for the estimation of the reliability of the method. However, the quality of the delivered data is not in all cases assured: the interpretation method of both tests (static and rapid) might differ between the cases, since several authors are involved. It is unclear whether or not the piles have mobilized ultimate behaviour. Recalculation of the results is not always possible, since the original measured data is not available. Without being able to verify these aspects it is difficult to assess the influence of the available interpretation methods and parameter selection.

The availability of well defined measurement data is an essential requirement for testing the described interpretation methods in detail. Also, the accuracy of newer analysis methods (for example the Sheffield method) can be tested, assuming the measurements are not used for improvement of the methods.

This Appendix describes the requirements and rules for usage of such a database. Since it was not possible to create a full database within the CUR/Delft Cluster project, a simplified version was created on internet. This is described in the final Section of this Appendix.

G.2 DATA FOR FURTHER ANALYSIS

A database set should contain two parts:

- A fully described rapid load test.
- A fully described static load test.

Further it must contain a description of the site, the piles, the installation and the tests carried out, together with the original measured data. Such a database set offers the possibility to judge the reliability of newly developed interpretation methods.

The description includes the measured data in digital form (ASCII) to make post processing possible.

G.3 DESCRIPTION OF DATABASE

The format for the cases in the database is clearly defined and the minimum requirements are given as well. All measured data must be available digitally, according to the definitions in the Standard. The requirements are less strict than those defined in the Standard, to make it possible to add older measurements and measurements with some minor shortcomings.

The minimum information required to accept a measurement for the database:

- A site description with layer thickness including basic soil classification information e.g. density index/moisture content, liquid limit and plastic limit, and a local in-situ measurement (e.g. SPT or CPT, in ASCII).
- A description of the pile(s): type, length, cross section and dimensions, material, stiffness. If the static test and the rapid test are not on the same pile, a remark why the piles are comparable. If the piles are not identical, a list with differences between the piles.
- A description of the installation: method, date of installation, toe level after installation.
- A description of the static test: date of execution, location of the pile, reaction method, working order, loading history of the pile.
  It is essential that treatment of creep during the SLT is described. The duration between load steps should be long enough that information on creep (or relaxation) is included.
- A description of the Rapid Load Test: date of execution, apparatus used, number of cycles, delay between cycles, location of the pile, loading history of the pile. The specifications of the device must be available, such as (reaction) mass, piston weight, specification of spring system. Additional specifications of each cycle might also be delivered: falling height, amount of fuel, observed displacement of the (reaction) mass, rebound of the mass.
- The measurement data for the static test (in ASCII).
- The measurement data for the rapid test (force, displacement, acceleration) as a function of time with proper sampling frequency (in ASCII). If the displacement or acceleration is not a measured signal, this must be stated.

The data will be reworked to the following documents:

- A report with information (points 1 to 4) as indicated above.
- An ASCII (or GEF-) file for the static test on each pile.
- An ASCII (or GEF-) file for each loading cycle of the rapid test on each pile.
- An ASCII (or GEF-) file with CPT or SPT data.

(note: GEF = geotechnical exchange file, see www.gef-files.org)

The report should contain graphs of the measured data, including a load-displacement diagram for each load cycle of the rapid load test, based on the measured data (without any correction).

If the static and rapid tests are carried out and reported according to the requirements of the Eurocodes, these requirements are met automatically. In that case, these two reports are published.

G.4 RIGHTS AND DISTRIBUTION OF RESULTS

The distribution of the content of the database will be free of charge. The entrance for downloading the data is free, but the data is password protected. This means registration is required and all users are known. By using the password, the user agrees with a general agreement.

The user agrees with the usage of an obligatory reference, which is part of the database. This reference refers to the original publication (prescribed by the owner of the data), and mention that the data are extracted from the database.

As an example: for the Delft Cluster measurements in Waddinxveen:

Report 413531-0053 "Demonstration project rapid pile load test Waddinxveen", Delft Cluster, Delft, the Netherlands, November 2008.
Data extracted from: Test database at www.rapidloadtesting.eu.

The first (two) lines are prescribed by the owner. The last line is common for all data.

For the time being, the data are not stored in a central database, but will be made available on request by the owner of the requested data.

G.5 CONTENTS OF THE CURRENT SIMPLIFIED DATABASE

At this moment the site www.rapidloadtesting.eu offers an overview of measurements which fulfil the requirements above and are freely available. The measurements are not distributed from the internet site, but one should request the data from the owner of the data, who is mentioned on the site. The owner can specify his own conditions, which might differ from the conditions suggested above.

Appendix H

Possible improvement of UPM for piles in clay

A theoretical estimation for the empirical factor for piles in clay can be obtained from the results in table 5.2. Based on Equation 5.7 in Section 5.3.1, the empirical factor η can be calculated at maximum force. Thus using for the quotient $F(t)/F_{max}$ the value 1. This gives

$$\eta = \frac{1}{1 + \alpha \left[\left(\frac{\tilde{v}(t)}{v_{ref}} \right)^{\beta} - \left(\frac{v_{static}}{v_{ref}} \right)^{\beta} \right]} \tag{H1}$$

Choosing for the velocity $\tilde{v}(t)$ the maximum velocity, an estimation of the empirical factor for the UPM is obtained. The Figure below gives (for all material properties in table 5.2) the empirical factor as a function of maximum velocity.

The maximum velocity during a RLT is about 1.0 m/s. This suggests that $\eta = 0.5$ is a reasonable average value for all soil types. However, due to deviation of the real velocity and depending on the material properties a value between 0.35 and 0.65 will be expected.

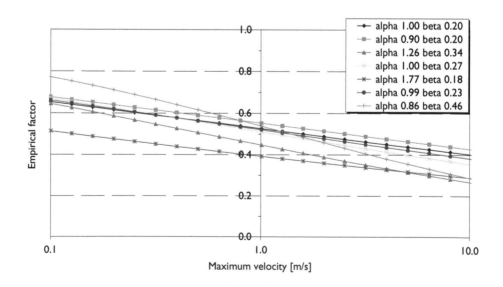

The mean value of 0.5 reasonably coincides with the mean value for clay mentioned in table 5.1. Using the Lognormal distribution and the data in table 5.1 for clay, the 10% lower limit for the empirical results is 0.22 and 90% upper limit is 0.90. So, the empirical observed variation is larger than estimated from the rate effect, but the usage of a velocity dependant empirical factor seems to increase the accuracy of the UPM for piles in clay. A similar approach is suggested by (Schmuker 2005).

This approach is not tested on the existing empirical data, so it must be used with care.

Appendix I

Draft standard for test execution

Draft standard

Geotechnical investigation and testing
Testing of geotechnical structures

Testing of piles: rapid load testing

reference EN-ISO 22477-##
date: 8 August 2010

I.I SCOPE

This standard establishes the specifications for the execution of rapid pile load tests in which a single pile is subject to an axial load in compression of intermediate duration to measure its load-displacement behaviour under rapid loading and an assessment of its static behaviour.

The provisions of this standard apply to piles loaded axially in compression.
This standard provides specifications for:

1   Investigation tests, whereby a sacrificial pile is loaded up to ULS (Ultimate Limit State).
2   Control tests, whereby the pile is loaded up to a specified load in excess of the SLS (Serviceability Limit State).

The magnitude of the rapid load tests must make allowance for the influence of rapid phenomena, such as rate effects, pore water pressures and inertia effects, when comparing with the equivalent static load test
Notes:

–   *Generally, an investigation test focuses on general knowledge of a pile type; a control test focuses on one specific application of a pile*
–   *While selecting the load in the rapid load test or during interpretation, one needs to take account of the effects in which the rapid load test differs from an equivalent static load test, such as rate effects, pore water pressures, creep and acceleration.*

1.2 NORMATIVE REFERENCES

The following referenced documents are indispensable for the application of this Standard:

EN 1990: 2002, Eurocode Basis of structural design
EN 1997-1, Eurocode 7: Geotechnical design – Part 1: General rules
NEN-EN-ISO 22477-1 Geotechnical investigation and testing – Testing of geotechnical structures – Part 1: Pile load test by static axially loaded compression

1.3 TERMS, DEFINITIONS AND SYMBOLS

1.3.1 Types of piles

Test pile: a pile to which loads are applied to determine the resistance and/or the deformation characteristics of the pile and the surrounding soil.

– Trial pile: a pile installed to asses the practicability and suitability of the construction method.
– Preliminary pile: a pile installed before the commencement of the main piling works or a specific part of the works for the purpose of establishing the suitability of the chosen type of pile and.
– For confirming its design, dimensions and bearing resistance.
– Working pile: a pile for the foundation of the structure.
– A test pile can be a trial pile, a preliminary pile or a working pile.

1.3.2 Rapid load

The load on the pile can be considered as a rapid load if the duration of the load fulfils the following

$$10 \prec \frac{T_f}{L/c_D} \leq 1000$$

The force shall be applied in a continuously increasing and continuously decreasing manner and should be sufficiently smooth.

1.3.3 Equivalent diameter

The equivalent diameter of a pile is:

– for a circular pile the outer diameter of the pile
– for a square pile the diameter which gives the same area as the square pile (as long as the longest side is smaller than 1.5 times the shortest side)
– for other piles the diameter which is used in static calculations of the pile

1.3.4 Reference distance

The reference distance is the distance which separates a stationary reference point from a point that will be significantly displaced by the testing method. Only stationary points can be used for reference of displacement measurements. Displacement measuring systems may be placed on the soil outside the reference distance without isolating (displacement compensating) measures.

The value of the reference distance is the maximum of:

– measured from the pile: the distance which the waves in the soil travel during the loading (T_f);
– measured from equipment with a falling mass: the distance which the waves in the soil travel during the falling of the mass and the subsequent loading (T_f).

1.3.5 Failure of a pile

Failure of a pile refers to geotechnical failure according EC 7 part 1.

1.3.6 Symbols

a acceleration [m/s^2]
c velocity of a stress wave [m/s]
c_p velocity of the stress wave in the test pile [m/s]
c_s velocity of the shear wave in the soil [m/s]
c_R velocity of the surface wave in the soil [m/s]
D (equivalent) diameter of the test pile [m]
e eccentricity [m]
f frequency [Hz]
F force [N]
g acceleration due to gravity [9.81 m/s^2]
L length of the test pile [m]
r distance [m]
r_{ref} reference distance [m]
T_f duration of the load [s]
w displacement [m]
v particle velocity [m/s]

Prefixes accepted by SI-units (such as [mm] instead of [m] and [kN] instead of [N]) are accepted.

1.4 EQUIPMENT

1.4.1 General

This standard is applicable to all types of equipment able to generate a loading at the pile head which fulfils the requirements in Section 3.2. If information on the ultimate bearing capacity of the pile is one of the goals of the test, the equipment must have enough capacity to reach the bearing capacity in rapid loading. The annex gives information on equipment which may fulfil the requirements.

Note: The required force on the pile head during a rapid load test for measuring the ultimate bearing capacity may exceed that for static loading by a factor of 2 in certain soil types.

If for a Rapid Load Test, one or more of the requirements mentioned in this Standard is not met it should be proven that this shortcoming has no influence on the achievement of the objectives of the test, before the results can be interpreted as a rapid load test.

I.4.2   Loading

The selection of the loading equipment shall take into account

– the aim of the test
– the ground conditions
– the maximum pile load
– the strength of the pile (material)
– the execution of the test
– safety aspects

The loading equipment must be able to generate a force which fulfils the requirements in 3.2 and generates the required maximum force.

If a test pile is tested by several cycles, the maximum force of each cycle should aim to be larger than the maximum force of the preceeding cycle.

The equipment must be able to load the pile accurately along the direction of the pile axis. The eccentricity of the load must be smaller than 10% of the equivalent diameter. The deviation of the alignment of the force to the axis of the pile must be smaller than 20 mm/m.

The stress in the pile under the maximum applied load must not exceed the permissible stress of the pile material.

Rebound of the mass on the pile head is not allowed without measurement of the resulting pile head load, deflection and acceleration.

I.4.3   Measurements

During a rapid load test three variables must be measured:

– the force applied to the pile head
– the displacement of the pile head
– the acceleration of the pile head

The transducers and signal processing must satisfy the following requirements.

Measurement	Requirement
general	
sample frequency	>4 kHz
duration of pre-event	>50 ms
duration of post-event	>300 ms
cut off frequency low pass filter	>1 kHz
load	
max. load	>target load
linearity	<2% of maximum value reached
hysteresis	<2% of maximum value reached
response time	<0.1 ms
acceleration	
number of transducers	≥1 for launching a mass, ≥2 for dropping mass
resonant frequency	>5 kHz
linearity	up to 50 g
displacement	
range	>50 mm and D/20
accuracy	±0.25 mm
response time	<0.1 ms
reference distance when launching mass	>15 m and c_s/T_f
reference distance when dropping mass	>15 m and c_s/T_f and *Stable* vibration free surface

Before and after each cycle, the level of the pile head must be determined relative to a point outside of the reference distance by high precision optical levelling.

The base of a displacement measuring system should not be placed within the reference distance (see I.3.4) from the pile. This must be verified at the test site. If the reference distance for a displacement measuring system cannot be reached, the displacement measuring system must be placed on a vibration free base.

The velocity of the pile head shall be calculated by integration of the measured accelerations with respect to time. Calculation of the displacement of the pile head by double integration of the measured accelerations with respect to time is allowed only if the final set is checked by a direct measurement of the displacement.

All equipment used for measuring load, displacement and acceleration in the test must be calibrated. The equipment must be checked on a regular basis. The results of these checks must be registered and kept with the most recent calibration. This data must be made available prior to commencement of the test.

Note: The time between the checks and calibrations is not prescribed, since the duration of validity of a calibration may depend strongly on the type of measurement device. However, the checks must be of that detail that one can be sure that all measurement devices are showing valid values during the test. It is preferred that all checks are carried out directly before the test, to avoid influence of transport and time.

All loadings (larger than 1% of the expected static bearing capacity of the pile) after installation of the pile must be measured. This includes all types of static preloading of the pile.

The measurement of the load on the pile head by strain gauges which are fixed to the pile, is allowed for internal forces only. The strain gauges must be calibrated against the results of an external load cell.

1.5 TEST PROCEDURE

1.5.1 Preparation

In advance of the test, an execution plan shall be formulated. The plan shall include the following:

1 Test objectives
2 Ground conditions, based on soil investigation in accordance with the regulations in EC 7
3 Testing date
4 Locations, types and specifications of the test piles
5 Allowable values of the load on the pile and the pile displacement
6 Required displacement of the pile and load on the pile
7 Specification of the loading device
8 Specifications of the measurement devices
9 Specifications of additional measurement-devices
10 Number of loading steps and target maximum load per step
11 Check on the acceptability of the foreseen load on the pile and displacement of the pile (with respect to the allowable values, defined item)
12 Duration of the measurement and sampling frequency
13 Plan of the test site
14 Time schedule
15 Personnel, showing who is responsible for

    a supervising
    b safety
    c loading
    d data recording
    e other tasks

16 Safety requirements
17 Legally required licences for handling the equipment
18 Other points of attention

1.5.2 Safety requirements

1.5.2.1 For people and equipment in the surrounding

Safety of human beings and equipment in the surrounding area must be assured during execution of the test. The distance between the nearest person and the test equipment must be at least twice the height of the test equipment measured from the soil surface.

People in neighbouring buildings must be informed about testing. Hindrance to vibration sensitive processes in neighbouring buildings shall be prevented.

I.5.2.2   *For the test pile*

The test pile should not be damaged by the test. During a rapid load test, the test pile will be loaded with a force which may exceed the static equivalent test loads by a factor of 2. Test piles should be designed to withstand the resulting higher stresses.

For working piles the maximum displacement of the pile head shall be agreed before commencement of the test. It may not exceed 10% of the (equivalent) diameter.

I.5.3   Preparation of the pile

The pile head must be flat, plane, perpendicular to the pile axis and undamaged.

The integrity and capacity of the pile must be sufficient to carry the planned test load. If installation of the pile causes doubts about pile integrity, the pile should be tested acoustically, or the rapid load test must be carried out by multiple steps with increasing pile load.

Note: Doubts about the integrity of a pile might be due to unexpected behaviour during construction. One might think of driving resistance, amount of concrete used, progress during drilling. The deviation might be a deviation from expected values or a deviation of a specific pile from other piles constructed at the construction site or similar site.

The test pile must have enough length above the ground surface to attach the measurement devices. All acceleration transducers must be installed firmly against or on the pile head.

Between the installation of the test pile and the beginning of the test, adequate time shall be allowed to ensure that the required strength of the pile material is achieved and the ground has sufficient time to recover from the process of pile installation and dissipation of pore-water pressures and other aspects, such as mechanical heat from boring or hardening concrete. The required waiting period may be assessed by measurements of e.g. excess pore-water pressure and soil strength evaluation. During this period, the pile may not be disturbed by load, impact or vibration, or other external influence.

The following time periods between installation and testing of a pile are prescribed:

– for trial and preliminary piles: minimum 7 days in non-cohesive soils, minimum 3 weeks for bored piles in cohesive soils and 5 weeks for driven piles in cohesive soils
– for working piles: minimum 5 days in non-cohesive soils, minimum 2 weeks for bored piles in cohesive soils and 3 weeks for driven piles in cohesive soils.

Note: These time periods are acc. ISO 22477-1 Sec 5.2.3 (draft).

I.5.4   General preparations

The sensitive parts of the test equipment shall be protected from weather (rain, wind, direct sunlight) and other disturbances.

All components of the system shall be protected against damage during all stages of construction and testing. Special attention must be paid to cables.

Any other site activities that might influence the measurements, e.g. vibrations by nearby traffic or ongoing pile driving, shall be avoided.

I.5.5 Aftercare of a working pile

If the result of the test causes doubts about pile integrity afterwards, the pile shall be tested acoustically.

Note: Doubts about the integrity of a pile might be due to unexpected behaviour during rapid load testing. The deviation might be a deviation from expected values or a deviation of a specific pile from other piles tested at the construction site or similar site.

I.6 TEST RESULTS

The test results shall consist of

- the force of the loading system [N] at the pile head as a function of time [s]
- the displacement of the pile head [m] as a function of time [s]
- the acceleration of the pile head [m/s^2] as a function of time [s]

There shall be a common base to all time measurements

All test results must be available in charts and digitally in an ascii-format. All results must be corrected for calibration factors. SI units are prescribed. Corrections applied to the measured signals must be put down in writing.

The measurements of pile levels before and after each cycle are reported. All other readings, such as temperature, tests on concrete samples, level readings, pile shape, static tests on the site, when relevant, must be put down in writing.

The rapid load-settlement diagram must be drawn. This diagram shows the measured pile head displacement [m] against the measured pile head force [N], without any correction.

A copy of all results shall be stored on a back-up medium.

I.7 TEST REPORT

The load test report shall at least comply with EN 1997-1. It should at least include the following information and data:

1   Reference to all relevant standards
2   General information concerning the test site and the test program:

    a   topographic location of the test
    b   description of the site
    c   purpose of the test
    d   test date
    e   the intended and realized testing program

    f   reference to the organization which carried out the test
    g   reference to the organization which supervised the test

3   Information concerning the ground conditions

    a   reference to the site investigation report
    b   location and reference number of the relevant soil tests
    c   description of the ground conditions, in particular at the vicinity of the test pile

4   Specifications concerning the test pile

    a   the pile type, its nomination and its reference number
    b   the topographic location of the test pile
    c   pile data, such as geometry, top and base level, pile material and reinforcement
    d   date of installation
    e   description of the pile installation and any observations related to the execution, likely to have an influence on the test results
    f   installation records, such as driving logs, concrete consumption, drilling progress

5   Specifications concerning the test

    a   the postulated maximum test load
    b   pile cap details
    c   details of the loading apparatus and measuring devices, including the calibration data
    d   the number of load cycles and the foreseen loading levels
    e   information on the potential energy for each cycle (drop height, mass, amount of fuel)
    f   the distance between the pile and the displacement measurement device
    g   details on the installation of the equipment by drawings and/or photographs

6   The test results

    a   as defined in chapter 6, including the digital data
    b   the rapid load-displacement diagram for each cycle from the measured signals
    c   the net settlement of each cycle
    d   the results of the high precision optical levelling

If the report includes an interpretation of the results with respect to the purpose of the test, the following information must be added:

a   the method used for the interpretation (with reference to the description)
b   the derived static load-displacement diagram
c   the treatment of rate effects
d   the treatment of mass effects
e   the treatment of effects due to excess pore pressure

APPENDIX I ANNEX A

Examples of equipment to which the standard is applicable Informative.

I-A1   STATNAMIC

The Statnamic Load Test (STN) has been developed by TNO and Berminghammer Foundation Equipment. The principle of the test is based on the launching of a reaction mass by burning fuel in a closed pressure chamber. This reaction mass is only 5% of the weight needed for a static load test. Loading is perfectly axial.

Figure I-A1 represents the successive stages of a Statnamic load test. Phase I is the situation just before launching. A cylinder with pressure chamber has been connected to the pile head and the reaction mass has been placed over the piston. In phase II the solid fuel propellant is ignited inside the pressure chamber, generating high-pressure gases and accelerating the reaction mass. At this stage the actual loading of the pile takes place, as an equal and opposite reaction force gently loads the pile. The applied pile force, displacement and acceleration are directly monitored. The upward movement of the reaction mass results in space, which is filled by the gravel (phase III). Gravity causes the gravel to flow over the pile head as a layer, catching the reaction mass and transferring impact forces to the subsoil (phase IV).

Available device loads are 1, 2, 3, 4, 5, 8, 16, 20, 30, and 40 MN. The testing range is between 25% and 100% of the device load. Devices of 100MN are under design and will be manufactured.

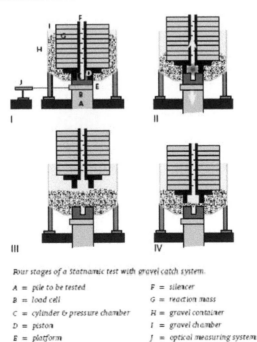

Four stages of a Statnamic test with gravel catch system.

A = pile to be tested   F = silencer
B = load cell   G = reaction mass
C = cylinder & pressure chamber H = gravel container
D = piston   I = gravel chamber
E = platform   J = optical measuring system

Figure I-A1 Stages of an Statnamic test.

During the test the reaction mass reaches a height between 2–3 m and then falls back. For high loads of 5–30 MN, gravel is used to catch the reaction mass. For loads in the range of 1–8 MN, a hydraulic catching system is utilized to arrest the reaction mass. With the latter system a considerable shorter cycle time is achieved, enabling more tests per day. With a 4 MN hydraulic catching device, 3–4 piles can be tested per day.

REFERENCE

Middendorp, P.; Bermingham, P.; Kuiper, B. Statnamic load testing of foundation piles. In: "Proc. 4th Int. Conf. Appl. Stress-Wave Theory to Piles, The Hague, Sept. 1992", Rotterdam, Balkema, 1992: 581–588.

I-A2   PSEUDO STATIC PILE LOAD TESTER

The load test by the Pseudo Static Pile Load Tester (PSPLT) is carried out by means of dropping a heavy mass (25.000 kg) with a coiled spring assembly from a predetermined height onto a single pile. After the hit, the mass bounces and is caught in its highest position. Catching the bouncing mass makes larger drop heights possible and avoids further hindrance to the test and the measurements.

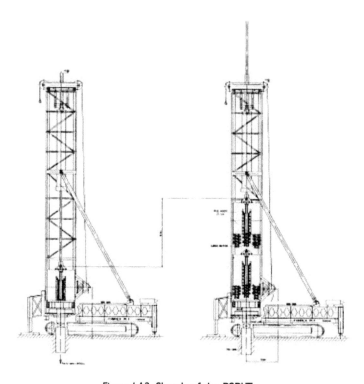

Figure I-A2 Sketch of the PSPLT.

The instrumentation for the test consists of a load cell and an optical displacement measuring device. The load cell is placed on top of the pile. It is almost identical to the one used during static load tests. Pile head displacement is recorded with the optical device mounted on a tripod at a distance of approx. 10 m from the pile. This tripod is equipped with a geophone to monitor vibrations of the tripod during the test. All measured signals are immediately processed by a computer and presented in relevant graphs.

The mass effects of the coiled springs in the PSPLT are minimized by using additional rubber springs and by creating a time delay between subsequent coils hitting the base plate. The spring stiffness is order 8 MN/m, but in fact a non-linear spring was installed.

The execution of a test is as follows: the PSPLT is brought to the test site by a low-loader. It moves on its tracks to the test pile, whose pile head has previously been prepared. When the rig is positioned and the measuring devices are attached the test starts. First a static load test is carried out with the weight of the drop mass. Then subsequently a number of rapid loads are deployed to the pile by dropping the mass from increasing heights onto the pile. With the output of results a quasi-static load-settlement curve is produced. Then the next pile can be tested. It is possible to load-test a significant number of piles per single working day. With proper preparations on the test site and the pile heads more than 10 piles daily have been tested.

REFERENCE

Schellingerhout, A.J.; Revoort, E. Pseudo static pile load tester. In: "Proc. 5th Int. Conf. Appl. Stress-Wave Theory to Piles, Orlando, Sept. 1996", Gainesville, Univ. Florida, Dep. Civ. Eng., 1996: 1031–1037.

I-A3 SPRING HAMMER TEST DEVICE

The loading mechanism of the Spring Hammer Test Device (SH device) is similar to that of the Pseudo-Static Pile Load Tester, except that the spring unit is placed on the pile head in the SH test. Two types of the SH device, portable and machine-mounted types, are available as shown in figure I-A3 and table I-A1. Coned disk springs are used to constitute a spring unit. The performance of the SH device can be easily controlled by changing the combination of the hammer mass and the spring value as well as the falling height of the hammer. One of advantages of the SH device is that repetitive loading can be easily done.

The applied force and the accelerations at the pile head are measured. Direct measurement of the displacement is possible by means of laser displacement meter or optical displacement meter. All the dynamic signals are recorded through a computerised signal acquisition system, and processed to estimate 'static' response of the test pile.

The SH device may be used very effectively to obtain the performance of piles having relatively low bearing capacity.

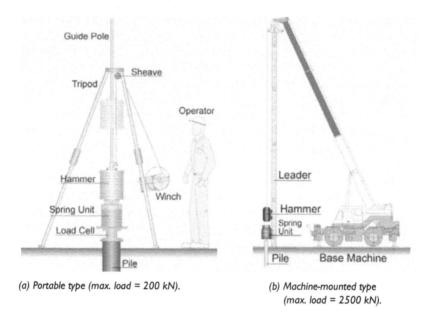

(a) Portable type (max. load = 200 kN).

*(b) Machine-mounted type
(max. load = 2500 kN).*

Figure I-A3 Spring hammer test device.

REFERENCE

Matsumoto, T., Wakisaka, T., Wang, F.W., Takeda, K. & Yabuuchi, N. Development of a rapid pile load test method using a falling mass attached with spring and damper, In: Proc. 7th Int. Conf. on the Appl. of Stress-Wave Theory to Piles, Selangor, Malaysia: 351–358.

APPENDIX I ANNEX B

I-B1   Interpretation of the test Informative

The measured force is not equal to the force which will be measured by at SLT. The test results must be interpreted

1   by comparing RLT and SLT at the test site
2   by using empirical relation, based on rapid load tests and static tests at sufficient comparable sites
3   an analytical or numerical model, validated at a sufficient number of load tests and static tests.

The following aspects must be evaluated during interpretation:

–   inertia effect
–   rate effect
–   generation of pore water pressure
–   plug-motion for open ended piles.

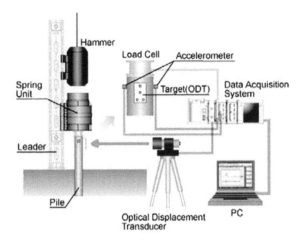

Figure I-A4 Signal acquisition system.

Table I-A1 Standard specifications of spring hammer devices (as in 2007).

	Portable	Machine mounted
Hammer mass (ton)	0.2	3
Spring values (kN/m)	5125	35000
Max. fall height (m)	2	3
Max. load (kN)	200	2500
Weight of spring unit (kN)	1	20
Number of tests per day in usual test condition	8 to 10	5 to 7

For the interpretation reference is made to a number of international documents.
An international guideline written by CUR, BRE, LCPC and WTCB. (to be published in 2011).
Additional information on the interpretation of the test can be found in
With respect to the inertia effect:

Middendorp, P.; Bermingham, P.; Kuiper, B. Statnamic load testing of foundation piles. In: "Proc. 4th Int. Conf. Appl. Stress-Wave Theory to Piles, The Hague, Sept. 1992", Rotterdam, Balkema, 1992: 581–588.

With respect to rate effects in clay:

Brown M.J.; Anderson W.F.; Hyde A.F. Statnamic testing of model piles in a clay calibration chamber In: Int. Jnl. Phys. Modelling Geotechnics., Vol. 4, No. 1:11–24 (ISSN 1346-213X).

With respect to the generation and influence of porewater pressure in sand

Huy, N.Q.; van Tol, A.F.; Hölscher, P. Interpretation of rapid pile load tests in sand in regard of rate effect and excess pore pressure to be published in Proceedings of the 8th International Conference of the Application of Stress-wave Theory to piles, Lisbon (PT), 8–10 September 2008.

With respect to the plug effects of open ended piles:

Ochiai Ochiai, H.; Kusakabe, O.; Sumi, K.; Matsumoto, T.; Nishimura, S. Dynamic and Statnamic load tests on offshore steel pipe piles with regard to failure mechanisms of pile-soil interfaces at external and internal shafts Proc. Int. Conf. on Foundation Failures, Singapore, 327–338, 1997.5.12–13.

APPENDIX I ANNEX C

I-C1   Information on requirements for the load
Informative

This annex shows an indicative method to judge the applicability of the applied force, which is defined in Section I.3.2. The method is shown graphically in figure I-C1. The requirements are defined in terms of the time derivative of the force.

Three times are defined:

t_{start} start of the loading
t_{max} the time the maximum load F_{max} is reached
t_{end} the end of the loading

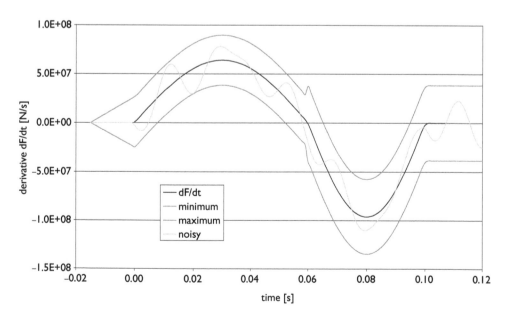

Figure I-C1 Calculated derivative of force.

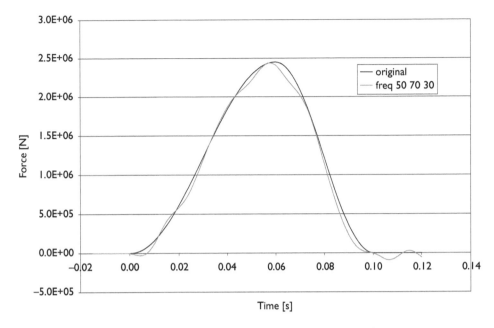

Figure I-C2 Force related to the derivative in figure I-CI.

The measured force is differentiated with respect to time. The resulting curve can be approximated by two half sine functions, one for the increasing part of the loading ($t_{start} < t < t_{max}$) and one for the unloading part ($t_{max} < t < t_{end}$). If the derivative does deviate less than 40% of the amplitude of the sine from the sine function, the loading can be accepted.

For loading equipment which has a different static force before and after the test (e.g. Statnamic starts with a non-zero load, a bouncing system might end with a non-zero load), the linear approximation must be corrected by a linear curve between the two static loadings at t_{start} and t_{end}.

Note: This method is a proposal to avoid discussion in future. Both the method and the limiting values must be tested against real measured data. Examples will be added.

APPENDIX I ANNEX D

I-DI   Information on requirements transducers and calibration Informative

Accelerometers
The application of servo electrical and piezo-electrical accelerometers is allowed. However, the integration of piezo electrical transducers with respect to time is more complicated. Therefore, it is advised to use servo-electrical transducers if the displacements must be calculated from the measured accelerations.

Moreover, the sample frequency must be higher if time integration is required.

National extensions

These guidelines refer to European and national documents. In this Appendix, the national documents are specified for each country.

Typical strain rate for SLT

Country	Document
Belgium	
France	
The Netherlands	MLT waiting 1–4 hours to permit creep of the soil >1 mm/s (NEN 6745)
United Kingdom	BSEN 1536, Special geotechnical works: Bored piles suggest 1 mm/minute for CRP

Predefined load displacement curve

Country	Document
Belgium	
France	
The Netherlands	The load in the calculated load-displacement curve can be scaled to the unloading point (NEN 6743)
United Kingdom	

National appendices

Country	Document
Belgium	
France	
The Netherlands	NEN-EN 1997-1/NB
United Kingdom	

National standard for static load test

Country	Document
Belgium	
France	
The Netherlands	NEN-6745-1 (nl) Geotechniek – Proefbelasten van funderingspalen – Deel 1: Statisch axiale belasting op druk
United Kingdom	

Author index

Subject index

Milton Keynes UK
Ingram Content Group UK Ltd.
UKHW051926141024
449569UK00027B/1371